# BETWEEN CRAFT AND SCIENCE

A VOLUME IN THE
COLLECTION ON

edited by **STEPHEN R. BARLEY**

# BETWEEN CRAFT AND SCIENCE

## Technical Work in U.S. Settings

EDITED BY

Stephen R. Barley
*and* Julian E. Orr

ILR Press | *an imprint of*

Cornell University Press

*Ithaca and London*

First published 1997 by Cornell University Press.
First printing, Cornell Paperbacks, 1997.

Printed in the United States of America

*Library of Congress Cataloging-in-Publication Data*

Between craft and science : technical work in U.S. settings / edited by Stephen R. Barley and Julian E. Orr.
p. cm. — (Collection on technology and work)
Includes index.
ISBN 0-8014-3296-0 (cloth : alk. paper). — ISBN 0-8014-8366-2 (pbk. : alk. paper)
1. Industrial technicians—United States. I. Barley, Stephen R. II. Orr, Julian E. (Julian Edgerton), 1945– . III. Series.
TA158.B47 1997
609.2'273—DC20 96-32077

This book is printed on Lyons Falls Turin Book, a paper that is totally chlorine-free and acid-free.

Cloth printing 10 9 8 7 6 5 4 3 2 1

Paperback printing 10 9 8 7 6 5 4 3 2 1

# CONTENTS

# PREFACE

Aside from the work of doctors, lawyers, and a handful of other visible professionals, relatively little is known about the content or the social organization of technical work. To be sure, science and engineering have attracted considerable attention over the years, but only recently have researchers begun to examine what scientists and engineers actually do and how their work is organized. Information on the work of technicians is even scarcer, although technicians now represent 3.6 percent of the American labor force and have been the fastest growing occupational category for several decades. The chapters in this book take a step toward remedying this situation.

Books are usually accretions of personal agendas, and this one is no exception. After studying technicians' work in several settings, we became convinced that the implications of the so-called shift to a postindustrial or service economy could not be fully understood without an appreciation of the expanding role that technical workers play in modern organizations. We thought progress toward such an understanding would be enhanced if researchers interested in technical work had an opportunity to pool their knowledge and lay the foundations of a research community. Although a small but growing number of scholars had become interested in technical work by the late 1980s, our perception was that many were unaware of each other's research because they spanned disciplines as diverse as sociology, anthropology, psychology, economics, engineering, and labor relations. We decided to remedy this situation. In 1991, under the auspices of the Program on Technology and Work at Cornell University's School of Industrial and Labor Relations, we embarked on the events that produced this volume.

In November 1992, with funding from the Department of Labor and Cornell's Institute for Labor Market Policy, we hosted a workshop at Cornell University on the technical labor force. The workshop brought together sixteen scholars from a variety of disciplines who had conducted field studies of technical work, but who were mostly unfamiliar with each other's findings. In addition to most of the authors whose chapters appear in this volume, Charles Goodwin, an anthropologist from the University of South Carolina, and Patricia Sachs, an anthropologist employed by NYNEX, also attended. The workshop's goals were to facilitate the sharing of information and ideas, to arrive at a working definition of technical work, to identify issues that all participants considered critical for future development, to facilitate collaboration, and, finally, to lay the foundation for this book and a second, somewhat larger, conference. During the workshop, the participants converged on a series of topics that they agreed deserved further elaboration. They then divided responsibility for developing papers on these topics. The commissioned papers clustered around four broad themes: (1) technical work's challenge to the established order, (2) detailed studies of technical work, knowledge, and practice, (3) detailed studies of technical workers' identities, values, and beliefs, and (4) training, credentialling, and careers in technical occupations. The goal was to develop papers that ranged coherently from the theoretical to the descriptive to the policy-oriented. This volume retains the organizational scheme developed at the workshop.

Between December 1992 and March 1994 the authors conducted additional research and produced initial drafts. These drafts were delivered at Cornell in March 1994 during a conference sponsored by the Alfred P. Sloan Foundation. The conference brought the authors together for three days with an audience composed of other researchers who have studied technical work as well as leaders from industry, labor, and government who are especially knowledgeable about issues pertaining to technical work and the technical labor force. After the conference, authors revised their papers in light of the dialogue that occurred.

This volume would not have been possible without the financial and intellectual support of Hirsch Cohen at the Alfred P. Sloan Foundation, Stephanie Swirsky of the Department of Labor's Employment and Training Administration, and Ronald Ehrenberg, Director of the Institute for Labor Market Policy. The researchers and organizers associated with Cornell's Program on Technology and Work are also deeply indebted to the U.S. Department of Education which, from the beginning, has funded our work on technicians through grants from the National Center for the Educational Quality of the Workforce.

Thanks are also due to Stacia Zabusky and Margaret Gleason, who organized the first workshop, to Renee Edelman, who orchestrated the conference, and to Paula Wright and Roger Kovalchick, who helped transform the papers into a single, integrated document.

STEPHEN R. BARLEY
JULIAN E. ORR

Palo Alto, California
October 11, 1995

# CONTRIBUTORS

**Lotte Bailyn**
Sloan School of Management
Massachusetts Institute of Technology
50 Memorial Drive
Cambridge, Mass. 02139

**Stephen R. Barley**
Industrial Engineering and Engineering Management
340 Terman Hall
Stanford University
Stanford, Calif. 94305

**Louis L. Bucciarelli**
School of Engineering
3-282
Massachusetts Institute of Technology
77 Massachusetts Avenue
Cambridge, Mass. 02139

**Larry Carbone**
School of Veterinary Medicine
Cornell University
Ithaca, N.Y. 14853

**H. M. Collins**
Centre for the Study of Knowledge, Expertise and Science
Dept. of Sociology and Social Policy
University of Southampton
Southampton SO17 1BS
U.K.

**Sean Creighton**
Department of Sociology
Ballantine Hall, Room 751
Indiana University
Bloomington, Ind. 47401

**Randy Hodson**
Department of Sociology
Ballantine Hall, Room 751
Indiana University
Bloomington, Ind. 47401

**Jeffrey Keefe**
School of Management and Labor Relations
Rutgers University
P.O. Box 231 Ryders Lane
New Brunswick, N.J. 08903

**Sarah Kuhn**
Policy and Planning Department
College of Management
University of Massachusetts–Lowell
One University Avenue
Lowell, Mass. 01853

**Bonalyn J. Nelsen**
Johnson Graduate School of Management
Cornell University
Ithaca, N.Y. 14853

**Julian E. Orr**
Xerox PARC
3333 Coyote Hill Road
Palo Alto, Calif. 94304

**Brian T. Pentland**
School of Labor and Industrial Relations
412 South Kedsie Hall
Michigan State University
East Lansing, Mich. 48824-1032

**Leslie Perlow**
University of Michigan Business School
701 Tappan Street
Ann Arbor, Mich. 48109-1234

**Trevor Pinch**
Science and Technology Studies
622 Clark Hall
Cornell University
Ithaca, N.Y. 14853

**Denise Potosky**
Graduate Program in Management
The Pennsylvania State University
Great Valley Campus
30 E. Swedesford Road
Malvern, Penn. 19355

**Mario Scarselletta**
Corning Inc.
Science Products Division
Corning, N.Y. 14831

**Peter Whalley**
Department of Sociology and Anthropology
Loyola University Chicago
Lake Shore Campus
6525 North Sheridan Road
Chicago, Ill. 60626

**Stacia E. Zabusky**
Associate in Research
Institute of European Studies
Cornell University
Ithaca, N.Y. 14853

# BETWEEN CRAFT AND SCIENCE

# INTRODUCTION: THE NEGLECTED WORKFORCE

*Stephen R. Barley and Julian E. Orr*

## THE CHANGING NATURE OF WORK

Work forms the bedrock of all economic systems. When the nature and social organization of work change, so does the fabric of society. On this dictum, Marx, Weber, Durkheim, and Tönnies anchored their analyses of the social transformation we now call the Industrial Revolution and, in the process, gave intellectual direction to the fledgling field of sociology. The Industrial Revolution transformed work in Western society in two related but analytically distinct ways. First and most obvious, it occasioned a massive shift in what people did for a living. Over the course of the nineteenth and early twentieth centuries, the balance of employment in every Western nation shifted from agriculture to manufacturing. Possibly more critical for the course of Western culture, however, was the second type of change: the shift in how people did their work, or what Marx referred to as a change in the "mode of production."

Most goods produced and consumed up to the late nineteenth century were not that different from those produced and consumed before the Industrial Revolution. What changed was how goods were made. Prior to the Industrial Revolution, textiles, plows, muskets, glassware, chairs, tables, and most other products were manufactured by hand with simple tools either in the home or in the shops of skilled artisans. The Industrial Revolution brought workers together in larger shops and factories, where propinquity allowed owners to divide tasks into constituent activities and assign those activities to individuals who performed them repeatedly. The advent of interchangeable parts and specialized machine tools furthered the progressive division of labor which, with time, became the distinctive signature of industrialization. By the early decades of the twentieth century, the shift in the nature of work was so far-reaching that "job" had come to mean employment in a vertically structured organization, and craft had been reduced to a secondary means of organizing work (Abbott 1989).

In the late 1960s, a handful of sociologists began to argue from several perspectives that the West was once again embroiled in an economic transformation whose scope would rival that of the Industrial Revolution. Their claim rested, in part, on the observation that the labor force was becoming increasingly white-collar. For Daniel Bell (1973) and other advocates of a "postindustrial economy" (Galbraith 1967; Touraine 1971), the transformation pivoted on the declining importance of manufacturing and the growing centrality of service industries that were heavily populated by "knowledge workers," whose elite consisted of scientists, technicians, and professionals. A number of French Marxists believed they saw in the same trends the rise of a "new working class." Because technical workers controlled the knowledge and technologies on which their expertise rested, Serge Mallet (1975) and André Gorz (1967) argued that the new working class would be more effective than the older working class at resisting the dynamics of capitalism. Still other sociologists argued that professional and technical workers actually represented a "new middle" rather than a "new working" class and that, therefore, their interests were largely aligned with the existing power structure (Poulantzas 1975; Wright 1978). All three schools of thought met with considerable skepticism, in part because each oversimplified the complexity of the changes it observed and in part because each postulated a shift in culture or social structure that was based more on wishful thinking than on evidence. Although the postindustrialists and the neo-Marxists may have overinterpreted the changes they observed, with each passing year it becomes more and more difficult to deny the accuracy of their most central observation: the occupational division of labor is again changing, and it is changing in an apparently consistent direction.

Trends in the United States since mid-century are illustrative. As Table I.1 indicates, even after a century of steady decline, agriculture still employed 12 percent of all Americans in 1950. By 1991, however, agriculture employed a mere 3 percent of the labor force. Considering it alone, one might view the continuing decline in agricultural employment as little more than the completion of a trend begun in the early years of the Industrial Revolution, but concomitant developments suggest a different scenario. During the Industrial Revolution, dwindling agricultural employment was offset by the expansion of semi-skilled and unskilled blue-collar labor and, somewhat later, by the expansion of the clerical and administrative workforce. Since mid-century, however, the number of unskilled and semi-skilled jobs has steadily declined. Whereas 26 percent of all Americans worked as operatives and laborers in 1950, only 15 percent of the workforce were

TABLE I.1
Occupational categories as a percentage of the labor force, 1950–1991

| *Category* | *1950* | *1960* | *1970* | *1980* | *1991* | *Net Change* |
|---|---|---|---|---|---|---|
| Farmworkers | 12% | 6% | 3% | 3% | 3% | −9% |
| Professional/Technical | 8 | 10 | 14 | 15 | 17 | 9 |
| Craft and kindred | 14 | 14 | 14 | 12 | 11 | −3 |
| Operatives/Laborers | 26 | 24 | 23 | 18 | 15 | −11 |
| Clerical and kindred | 12 | 15 | 18 | 17 | 16 | 4 |
| Service | 11 | 12 | 13 | 13 | 14 | 3 |
| Managerial/Administrative | 9 | 8 | 8 | 10 | 13 | 4 |
| Sales workers | 7 | 7 | 7 | 11 | 12 | 5 |

*Note:* Percentage employment by occupational category from 1950 to 1970 was calculated from employment data presented on page 139 of *The Statistical History of the United States from Colonial Times to the Present* (U.S. Bureau of the Census, 1976). Data for 1980 were taken from Klein's (1984) article which transforms 1980 data using the Census Bureau's category system developed in 1983. Data for 1991 are taken from the *Statistical Abstract of the United States* (U.S. Bureau of Commerce, 1991).

so employed by 1991.[1] Over the same period, the proportion of Americans employed in skilled crafts also declined from 14 to 11 percent. The demise of blue-collar work is hardly news. Less well recognized is that clerical work has also begun to wane. Clerical employment in the United States peaked in 1970 at 18 percent of the workforce. By 1991, the proportion of Americans employed as clerical workers had fallen to 16 percent.

The upshot of these developments, as almost everyone knows, is that work has become increasingly white-collar and oriented to the provision of services. However, as Table I.1 makes clear, the nature of the change is not what many discussions of the service economy imply. For instance, although the proportion of Americans classified as service workers has grown since 1950, the 3 percent increase (from 11 to 14 percent) is actually smaller than the increases shown by all other occupational clusters that grew during the same period. Thus, we submit that if the data in Table I.1 indicate a shift toward a service economy, it does not appear to be an economy dominated by the low-wage, low-skill jobs that the government classifies as service occupations. Instead, Table I.1 suggests that professional and technical jobs are increasing faster than all others and may become the modal form of work for the twenty-first century.

The number of professional and technical jobs in the United States has grown by more than 300 percent since 1950 (see Figure I.1). No other occupational sector has experienced nearly as great a growth rate. Even sales (248 percent) and

1. The decline of unskilled and semi-skilled blue-collar labor actually began after 1940, when the proportion of Americans employed as operatives and laborers peaked at 28 percent of the labor force.

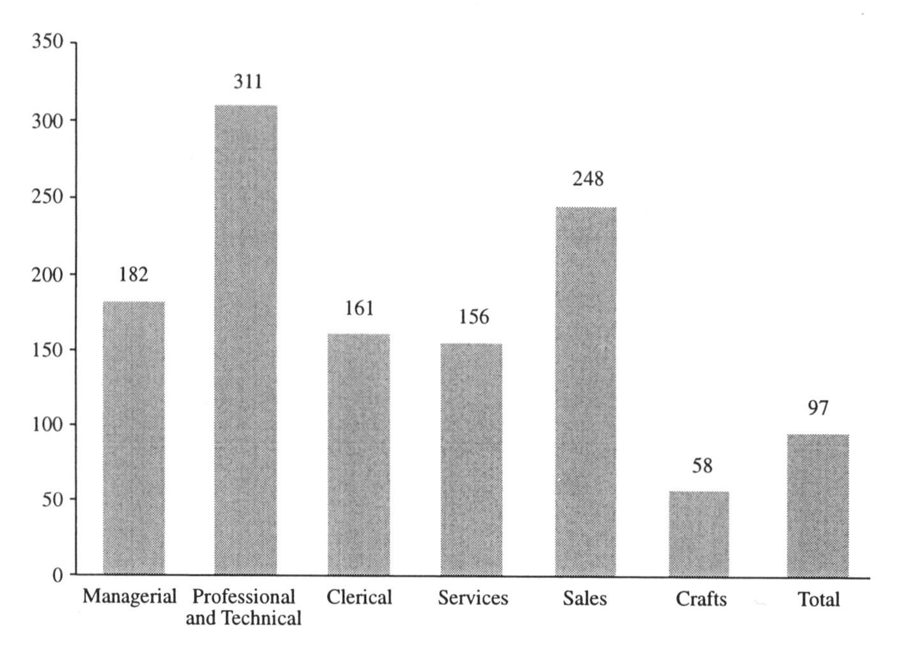


FIGURE I.1
Percentage growth in U.S. occupational categories, 1950–1991

managerial work (182 percent), which expanded tremendously over the last two decades, lag behind the growth of the professional and technical labor force. A quarter of all new jobs currently being created in the United States are either professional or technical in nature (Silvestri and Lucasiewicz 1991). As Table I.1 shows, by 1991 more Americans were found in professional and technical occupations (17 percent) than any other occupational category tracked by the Bureau of the Census. The Bureau of Labor Statistics' most recent employment forecast indicates that professional and technical workers will represent 18 percent of the labor force by 2005 (Silvestri and Lucasiewicz 1991). Similar trends are to be found in employment data from Canada and Great Britain, where professional and technical workers already account for 18 percent and 20 percent of the workforce respectively (Barley 1995).

Sociologists have long criticized aggregate occupational data for a variety of reasons, including the fact that government categories lump together occupations in an analytically naive, if not haphazard, way (Spenner 1980; Miller et al. 1980). This creates the possibility of either overestimating or underestimating the importance of change in any category; however, there is good reason to believe that in the case of professional and technical occupations, classification problems are likely to lead to conservative estimates of growth. Whereas most of the occupations classified as professional or technical by the U.S. government also satisfy sociological and cultural criteria for the label, numerous occupations that the government currently allocates to other categories also have strong technical components. For instance, the census classifies individuals who repair and maintain computers as craftspersons, even though they are typically called technicians in everyday life. Accountants are labeled managers, although sociologists of work and accountants themselves generally consider accounting to be a profession. Computer operators are classified as clerical and kindred workers. A sense of the extent to which census categories may undercount professional and technical workers can be gleaned from the fact that if classified differently, these three lines of work alone would raise the proportion of Americans classified as professionals and technicians by a full percentage point.[2]

If problems of "misclassification" were suddenly resolved, aggregate occupational data would still provide a hazy indicator of the shifting nature of work because, by definition, such schemes are based on job titles. At best, they index changes in what people do, but they are largely insensitive to changes in how people do what they do, unless shifts in technique also lead to changes in occupational nomenclature. Aggregate occupational data, therefore, are generally blind even to systematic changes in the way work is performed. Yet, as we have noted, the Industrial Revolution merits its name not simply because people began to pursue different lines of work, but because people also began to do old lines of work in radically new ways. Although it is difficult to gauge the extent to which qualitative changes are occurring in the nature of work, mounting evidence indicates that such change may be widespread. Moreover, it appears that the change points in a consistent direction, toward what might be called, for lack of a better term, the "technization of work."

By technization, we mean to characterize the emergence of work which is comparatively complex, analytic, and even abstract, because it makes use of tools that generate symbolic representations of physical phenomena and that often mediate between workers and the objects of their work. Controlling a nuclear power plant though an array of computer terminals is one example of such work; manipulating and studying the properties of cells using a cytometer is another. In some cases, technization describes the growth of new occupations. The work of sonographers and CT technicians, for example, arose *de novo* during the 1970s as hospitals added ultrasound and CT scanners to their arsenal of medical imaging devices. In other instances, technization proceeds by transforming existing lines of work. The second process is usually less visible than the first because it

2. Computer and electronics repair employed 0.1 percent of all working Americans in 1991. Another 0.1 percent of the population were employed as computer operators. Accountants represented 0.8 percent of the workforce (Silvestri and Lucasiewicz 1991).

does not change what an occupation is called and because practitioners appear to outsiders to be doing what they have always done, even though they now do it in dramatically different ways. The nature, implications, and potential scope of the second form of technization are most easily illustrated by examples drawn from lines of work that few consider technical: farming and the operation of continuous process plants.

A growing number of dairy farmers are beginning to rely on computer systems to manage their herds (Price 1993). As well as using computers to keep track of a farm's finances, dairy farmers can now monitor their cattle with electronic ear tags that identify individual cows. The tags work in concert with a coordinated system of sensors to generate data on how much milk each cow provides per milking, the rate of flow, how long each milking lasts, and the amount of feed a cow consumes over the course of a day. The data is used to make decisions about the cow's diet. When a cow's milk production falls, the tag issues a warning so the farmer may summon a veterinarian to examine the animal's health. In still other applications, sensors in fields send data on soil moisture and atmospheric humidity to a computer that combines these data with data from recent weather forecasts to control irrigation. Such systems are reputed to save both water and energy while producing healthier crops and more abundant yields (Smith 1984).

Operators in continuous process plants used to rely on sight, sound, smell, taste, and feel in addition to thermometers and pressure gauges to monitor production runs. As Shoshana Zuboff (1988) described in her influential study of pulp paper mills, operators now sit in elevated, air-conditioned control rooms where they monitor and intervene in production using an array of workstations that display and process data collected from hundreds of sensors spread across the factory floor. As a result, the operator's job has become increasingly analytical. Similar control technologies are rapidly altering production in other continuous process plants whose products range from peanut butter to petrochemicals.

## ENGINES OF CHANGE

### *The Growth and Commercialization of Scientific Knowledge*

Several interwoven dynamics appear to have occasioned both the technization of work and the expansion of the professional and technical labor force. Of paramount importance has been the harnessing of science for commercial advantage. Over the last century, industry has come to view science as a virtually limitless source of ideas for new products, new production processes, and even new industries. The commercialization of chemistry and physics during the last two decades of the nineteenth and the early decades of the twentieth century gave rise to the industries on which most Western economies now pivot: aerospace, energy, pharmaceuticals, petrochemicals, and electronics. Advances in the life sciences, especially in immunology, microbiology, biophysics, and biochemistry, made possible the expansion of the health care industry after World War II. More recently, molecular biology and computer science have created opportunities for entirely new industries and have revolutionized others.

The growth and commercialization of science has shifted the balance of employment toward technical work in three ways. First, the escalating demand for scientific knowledge, both basic and applied, has translated directly into employment opportunities for scientists, engineers, and other technical professionals. As the chemical and biotechnology industries grew, for example, so did their respective demands for chemists and molecular biologists. The demand for scientific and technical talent has done more than foster employment for members of existing disciplines, however; it has also triggered a proliferation of new fields and occupations. As a technical discipline grows, it becomes increasingly difficult for individuals to master the breadth of knowledge necessary to remain a generalist. Generalists are also less well prepared than specialists to provide "state-of-the-art" services. Consequently, most sciences and professions divide themselves into ever narrower subfields as their knowledge base grows. All else being equal, specialization should increase the number of employed professionals by requiring collaboration. Few specialists can execute alone tasks that require both breadth and depth of experience. Thus, as specialization proceeds, the number of experts necessary to accomplish a complex task, such as rendering a medical diagnosis, burgeons. The growth of scientific knowledge spawned technical work by yet a third path. As fields grow, scientists and professionals tend to allocate more routine duties to somewhat less well trained individuals. As Peter Whalley and Stephen Barley note in Chapter 1, numerous technicians' occupations are rooted in precisely such a "hiving-off" process. Thus, scientific specialization encourages the proliferation of secondary support occupations which are themselves technical.

### *Technological Change*

Another critical factor in the shift to a technical workforce has been technological change. Throughout history, technologies have spawned new occupations: the wheelwright, the blacksmith, the machinist, the automobile mechanic, and the airline pilot are illustrations. In the past, technologies created occupations across the entire division of labor. Although modern technologies have also produced occupations in all strata, those with high technical content appear to have become more common (Adler 1992). The advent of computer technology and the subsequent shift from mechanical to microelectronic devices are largely responsible for the difference.

In 1950, for instance, few people worked with computers and most who did were mathematicians (Pettigrew 1973). By the 1970s, computers had given birth to such well-known occupations as programmer, systems analyst, operations researcher, computer operator, and computer repair technician. These occupations continue to be among the fastest growing. In the United States alone they are anticipated to provide employment for 2.3 million people by the turn of the century, or 1.6 percent of the American labor force.[3] The explosion of occupations directly related to the computer, however, is only the most visible sign that technology may now favor a technical labor force. Numerous technical occupations have been created over the last four decades by technologies other than the computer; nuclear technicians, nuclear medical technicians, broadcast engineers, and materials scientists are examples. Moreover, computers have altered the contours of many traditional jobs.

To grasp how computers have accelerated the technization of existing work, one must distinguish between substitutional and infrastructural technological change (Barley 1991). Most technological change is perceived to be a matter of substitution, replacing an earlier technology with a more efficient or effective successor. However, the simplicity of substitution is usually illusory. Work with a new technology will rarely be done in the same way as it was with its predecessor, and so the changes will necessarily be more complex than is suggested by substitution. Nevertheless, historically, substitute technologies have usually been adopted on the grounds that they make some parts of work easier to perform. Some have also generated considerable profits by reducing labor costs and allowing economies of scale. By comparison, infrastructural change is rare. Infrastructural technologies are the relatively small set of technologies that form the bedrock of a society's system of production during a particular historical era. Until recently, the economies of the industrial nations revolved around electrical power, the electric motor, the internal combustion engine, and the telephone (Coombs 1984). The diffusion of these technologies occasioned the Second Industrial Revolution.

New technologies, even those that alter the infrastructure of production, are initially perceived as substitutes for existing technologies, and microelectronics have been no exception. Early computers were adopted by organizations to streamline personnel, accounting, and other paper processing operations. Similar motives have underwritten the spread of most other digital technologies. Digital sensors, robots, production scheduling systems, and computer-integrated

3. Estimates are based on data from Silvestri and Lucasiewicz's (1991) estimates for a moderate growth scenario.

machine tools, for instance, purport to make manufacturing processes more efficient by replacing slower mechanical devices, paper-based inventory systems, and semi-skilled labor. Office automation enabled firms to reduce the number of clerical personnel they employ, reputedly by allowing offices to increase the volume of data they process per unit time. Because digital technologies are often developed, bought, and sold explicitly for their substitutional promise, early sociological research on computers and other microelectronic technologies focused almost entirely on reductions in force and the deskilling of work (Zimbalist 1979; Crompton and Jones 1984; Noble 1984).

Yet, even in their purely substitutional role, microelectronic technologies have furthered the shift to an increasingly technical labor force. In addition to requiring new skills of their users, these technologies require a cadre of skilled technicians capable of programming, diagnosing, and repairing devices. Barbara Baran (1987), Thomas Diprete (1988), and Paul Attewell (1987) have shown that the movement of computers into the insurance and banking industries led to a slight increase in average levels of skill primarily because the shift to computational technologies required firms to hire programmers and technicians even as they reduced the number of clerical employees. Harry Braverman (1974), who first articulated the deskilling thesis, also recognized that firms would require a cadre of programmers and technicians if they were to use numerical control successfully to deskill or replace machinists. Only recently, however, have students of technology begun to realize that microelectronics' greatest influence on the social organization of work may arise from its infrastructural rather than its substitutional potential (Adler 1992, Barley 1991). It is as an infrastructural technology that microelectronics becomes implicated in the technization of existing lines of work. Even in cases such as clerical work where the deployment of microelectronic technology has been perceived as pure substitution, the need to cope with operating systems, servers, networks, and applications has clearly wrought a fundamental change in daily tasks.

The link between a microelectronic infrastructure and the technization of work rests on the concept of the digitization of information. Digitization has occurred in two distinct waves. The first centered on the translation of symbols into digital impulses that could be stored, manipulated, transferred, and decoded in electronic form. It is this type of digitization that underlies the most familiar uses of computational technology: word processing, numerical calculation, database management, and electronic mail. The computing infrastructure created by the technologies associated with the first wave of digitization made it possible to traffic in large datasets. The ability to amass and analyze huge databases, in turn, gave cultural credence to the notion of information as an economic good. Data were equated with information, a commodity that could be created, bought, and sold. The commodification of data as information accounts for the growth of a number of industries and professional occupations that are neither directly dependent on the expansion of scientific knowledge nor directly focused on the maintenance of the infrastructure itself. Examples include the financial services industry, commercial database services, and a variety of consulting services.

The second and more recent wave of digitization has brought sophisticated sensors that transform physical signals into data and effector technologies that use digitized data to control machines and manipulate objects. Integrated sensing and manipulating technologies have been critical to the rise of computer-integrated manufacturing and the advent of new medical imaging devices. In the long run, the second wave of digitization may prove to be more critical for the restructuring of work than the first. The ability to convert physical signals into data that can be used to manipulate objects opens the possibility for radically new modes and relations of production. Not only do such technologies promise to reduce even more dramatically the number of humans directly involved in manufacturing, but they are likely to engender forms of work in which workers are increasingly distanced by digital interfaces from direct contact with the physical systems over which they have responsibility. Here, then, lies the essence of the technization of work: the increasing ability to intervene in the world of objects through symbols. The work of pulp paper mill operators described by Zuboff exemplifies such technization. Whereas the mode of production associated with the Industrial Revolution revolved around the production and manipulation of things, the mode of production associated with the shift to a professional and technical labor force is likely to revolve around the creation and manipulation of symbols in order to work on things.[4]

## WHAT IS TECHNICAL WORK?

Far more difficult than pointing to the expansion of technical work is the task of defining what technical work is and how it differs from other types of work. One common definition draws credibility from the claim that technological

4. Although the growing economic importance of scientific knowledge and infrastructural technological change are the primary reasons for the burgeoning of technical work, it is important to recognize that both factors operate in concert with other social trends. As a case in point, consider the recent phenomenon euphemistically known as "downsizing." Although downsizing has not directly created professional and technical jobs, it has enhanced their prominence by disproportionately targeting blue-collar and mid-level managerial jobs for elimination. As a matter of accounting, the relative prominence of professional and technical work should increase as other occupational categories shrink. Although some commentators have argued that widespread downsizing has become possible largely because of computer technology, such accounts are suspect because computers have not yet significantly automated the work of middle managers. At best, downsizing appears to be a fashion whose unintended consequence has been to increase the complexity of remaining jobs and to enhance the prominence of the technical and professional workforce, which has so far largely escaped wholesale reductions.

change is responsible for the emergence of the technical labor force. From this perspective, technical work entails either working "on" a complex technology, as in the case of repair, or working "through" or "with" a complex technology to achieve some other end. This definition is attractive because it enables one to count as technical not only the work of technicians who maintain equipment, but also work that has been technized. One drawback is that the definition provides no clear criteria for distinguishing between traditional blue-collar workers who simply operate complex machinery and those whose function has been fundamentally transformed. The principal weakness of the definition is that it does not address what is actually done at work.

Another way of distinguishing technical work from other lines of work is focusing on how members of different occupations know what they know. From this perspective, technical work would be distinguished by the type of knowledge it requires. Perhaps the most common definition of technical work employs this approach: Technical work is work that requires one to understand and utilize an abstract body of knowledge. Those who adopt such an approach usually claim that technical work requires familiarity with mathematics or science, an appreciation of abstract principles that underwrite practice, or the mastery of a way of knowing that is somehow more "objective" than other ways of knowing (Barlow 1967; Hull 1986). Analysts who adopt such a definition tend to associate technical work with analytic tasks and to focus on formal education as a source of training. The advantage of such a definition is that it allows one to treat the work of most professions as technical work. The disadvantage is that the focus on formal knowledge, rather than practice, excludes lines of work whose practitioners rely almost exclusively on contextual knowledge, but who are considered by almost everyone to be technical workers, for example, photocopier technicians (Orr 1996), microcomputer support technicians (Zabusky, Chapter 6), and almost anyone else whose job centers on the repair of complex technologies. More important, the approach devalues the fact that most studies of technical work have repeatedly shown that technical acumen, even among professionals, draws heavily on contextual and even tacit knowledge (Collins 1974; Knorr-Cetina 1981; Lynch 1985).

A third approach to defining technical work begins with the observation that all work has a technical component. In everyday speech, we often use the term "technical" to refer to skilled practice, the use of technique to accomplish work. This everyday meaning is accurate, but for our purposes it is too inclusive. In some sense all work qualifies, and under this interpretation technical work differs from other forms of work only in degree, not in kind. Furthermore, this meaning includes only practice and does not recognize the abstract or formal knowledge which typically seems to be part of technical work.

A critical aspect of our everyday understanding of technical work is that it is somehow qualitatively different from other types of work. Technical work sits at the intersection of craft and science, combining attributes of each that are normally thought to be incompatible. It is a cultural anomaly in which mental and manual skills coexist inseparably, if not always comfortably. For this reason, none of the previous definitions are adequate. Therefore, we submit that it is more productive to identify technical work by a loose constellation of attributes, several of which are not normally juxtaposed in traditional frameworks for classifying work. Although individual instances of technical work may fail to evince all characteristics, all cases will possess the majority. Four traits comprise the constellation we propose: (a) the centrality of complex technology to the work, (b) the importance of contextual knowledge and skill, (c) the importance of theories or abstract representations of phenomena, and (d) the existence of a community of practice that serves as a distributed repository for knowledge of relevance to practitioners.

This understanding of technical work offers several advantages over the previous approaches. It enables the analyst to count as technical the entire array of occupations that fit common language definitions of technical work. Specifically, it allows us to treat most professionals and technicians as technical workers, as well as all those whose work has been technized, while excluding most managers as well as the incumbents of *traditional* clerical and blue-collar jobs that are better understood within an industrial framework. More important, it highlights the fact that technical work blurs the traditional boundaries between professions and crafts, and between blue- and white-collar jobs. Categories of technical work are anomalous in this combination of skill and abstract knowledge and often seem to fall into the gaps between better-known categories. Veterinary surgeons, for example, are universally acknowledged as professionals, but they work with animals and depend on manual skills. Their work, skillful cutting, is craft, but their profession is medicine. Emergency medical technicians work in the gap between the medical profession and a service occupation. They are connected to the first by a radio, their skills, and a working grasp of medicine; to the second, by their role as ambulance drivers. Engineers work between production workers and managers, on the one hand, and between craft and science on the other. Technicians, perhaps, best exemplify the anomalous status of technical work. They work as craftspeople informed by abstract knowledge, and their standing in society reflects the interstitial location of their work, as does the very etymology of the term "technician."

### *Technicians*

The word "technician" appears to have entered the English language from French during the first third of the nineteenth century as a synonym for the then more commonly used "technicist." Although derived from Greek and Latin roots that denote craftwork (Partridge 1966), "technician" initially indicated skill but had no occupational connotations and no link to science or technology. It was a term applied to artists who had mastered the technicalities of their medium. To be a technician in the nineteenth century, according to the *Oxford English Dictionary,* was to be "skilled in the technique or mechanical part of an art, as music or painting." Although being called a technician acknowledged skill, the term was not wholly flattering. The frequent juxtaposition of "technician" with the adjective "mere" implied a competent practitioner who nevertheless lacked artistic gifts. Real artists were born; technicians were schooled. From the beginning, technicians appear to have had ambiguous status: to be called a technician was to be simultaneously respected and denigrated.

By the early 1900s, "technician" had acquired occupational overtones. Most dictionaries at the turn of the century defined technicians as individuals versed in the technical aspects of any subject, including the "practical arts," a nineteenth-century term for craft with an industrial flavor. The modern sense of "technician"—a person whose work revolves around instruments and who may require specialized training in a science or technology—did not appear until after World War II. The definition of the term in the 1947 edition of *Funk and Wagnall's* is particularly enlightening:

> (1) One skilled in the handling of instruments or in the performance of tasks requiring specialized training. (2) A rating in the armed services including those qualified for technical work; also, one having such a rating.

The second entry indicates that the military may have been the first organization to use the term "technician" to denote a specific type of occupation. The military usage apparently diffused into the larger culture after World War II. Other dictionaries from the 1940s and 1950s support such an inference (e.g., *Webster's New World Dictionary* 1959). By the 1960s, references to a Military Occupational Specialty had disappeared from dictionaries, although the number of servicemen and servicewomen called technicians had increased. The fact that those who compiled dictionaries no longer felt obliged to distinguish between common and military usage suggests that the military's sense of the term had become colloquial. Thus, the *Oxford English Dictionary*'s most recent (1989) definition of technician simply reads:

> A person qualified in the practical application of one of the sciences or mechanical arts; now esp., a person whose job it is to carry out practical work in a laboratory or to give assistance with technical equipment.

The trajectory by which "technician" became the name for an occupational category suggests why technicians' work may be emblematic not only of the socioeconomic factors that have led to a more technical labor force, but also of the cultural dilemmas posed by such a labor force. Technicians' work is simultaneously associated with science and craft, which have historically stood on opposite sides of the divide between mental and manual labor. The advent of technicians' work accompanied the rise of electronic technologies, the growing economic importance of science, and, by implication, the decline of the traditional industrial regime. Perhaps most intriguing, technicians seem to have always occupied a culturally ambivalent station: to be a technician is to be highly skilled and yet to be called a technician is to be subtly maligned. We propose that this difficulty arises, at least in part, because technicians mediate between technology and society.

Technicians can be said to stand between technology and society in two senses, the first cultural, the second structural. They are culturally situated between technology and society because technological change has shoved them toward the economic heart of a society not quite ready to leave behind the categories of industrialism, its distribution of power, and its presumed distribution of knowledge. Technicians stand somewhere between our future and our past, confronting us with contradictions that confuse and, at times, threaten us. As Stacia Zabusky notes in Chapter 6, technicians are expected to be servants as well as experts. They are expected to ensure that the "system" runs and to rescue us from the complexities and "normal accidents" (Perrow 1984) of the technologies we create but no longer understand. Yet, at all other times, technicians are supposed to be invisible, perhaps to support the pretense that nothing has changed and the world is still as it was.

Technicians stand between technology and society in a structural sense as well, and it is here that their power lurks. They link us to technologies that are nearly transparent when they work and troublesomely opaque when they do not. They enable us to get on with our lives without knowing too much about the machinery that runs in the background but makes our lives, as we know them, possible. In this sense, then, technicians separate us from the technology on which our society is based. Barley (1993) has argued that differences in the way this mediation is structured underwrite a typology of technicians' work. Some technicians, such as science technicians, engineering technicians, sonographers, and emergency medical technicians, work as "buffers": They simultaneously link professionals to, and shield them from, aspects of the material world about which those professionals are presumably experts. In one way or another, such technicians reduce material phenomena to information that becomes grist for the mill of another occupation. Other technicians serve as "brokers": they link nontechnical communities that use technologies to the technical communities that produce them. Often brokers adapt the technologies to the needs of the user community and ensure that the system remains operational. Examples of brokers include computer technicians, photocopier technicians, and automobile technicians. Brokers and buffers occupy a critical role precisely because they stand between the technology in which they are expert and the segment of society that the technology serves. As Michel Crozier (1963) observed with respect to mechanics, such a position often brings latent power incommensurate with one's status. The technicians' power rests on the fact that removing them from a production system would quickly lead to the system's collapse.

For this reason, technicians are paradigmatic of all technical workers. In its various forms, technical work has become socially essential precisely because it allows all other types of work to proceed. It has become the fulcrum of the emerging economy, even though technicians remain the most neglected members of the workforce. We believe that the time has come to examine technical work on its own terms and to ask whether it challenges modes and relations of production that have gone before, and, if so, how. Because the cluster of attributes that define technical work depend on the doing of the work, we think the most appropriate strategy is to study its practice. This collection of papers represents our undertaking of that effort.

## DISCONTINUITIES AND DISJUNCTURES: FROM THEORY TO PRACTICE TO POLICY

The ten chapters that follow were written explicitly to take the reader on a three-legged intellectual journey. The papers in Part I comprise the first leg of the journey, exploring the position of technical workers in modern society. They highlight the cultural ambiguities that surround technical work and point to the inadequacy of existing theories for making sense of the technical worker's position in the division of labor. Peter Whalley and Stephen Barley detail the anomalous position of the technical workforce by showing that engineers and technicians violate long-standing distinctions on which our notions of work are based. In particular, Whalley and Barley argue that technical work poses a cultural problem because it bridges the chasm between mental and manual work and belies the assumption that work must be structured according to either organizational or occupational principles. These anomalies upset the established order, causing problems of control for organizations and problems of verisimilitude for social theorists.

The papers by Jeffrey Keefe and Denise Potosky and Sean Creighton and Randy Hodson further substantiate the notion that technical work is socially anomalous. Together the two papers show that technicians do not quite resemble the members of any other occupational group or, for that matter, behave in the ways social theories say they should. Focusing on aggregate occupational data as well as a case study of technicians in a pharmaceutical company, Keefe and

Potosky demonstrate that technicians are in some ways similar to professionals, in other ways similar to craftspeople, and in yet other ways similar to factory operatives. Most intriguing is their finding that technicians are well-educated, relatively well-paid, and among the most highly skilled employees, yet they express more job dissatisfaction than any other occupational group, including semiskilled and unskilled laborers. Using survey data, Creighton and Hodson take a different theoretical stance but reach conclusions that parallel Keefe and Potosky's. Although technical workers resemble professionals in that they have high degrees of skill and enjoy reasonable prestige, their attitudes toward management are more like those of blue-collar workers. On the basis of these patterns, Creighton and Hodson suggest that technical workers be considered a "distinct category," not well described by existing sociological theory. The papers in Part I explain why we have little choice but to study what technical workers do, if we wish to understand the social organization of technical work.

In Part II the reader journeys to the level of detail. Each paper in Part II focuses a trained ethnographic eye on the work and culture of a single technical occupation. Trevor Pinch, H. M. Collins, and Larry Carbone analyze how veterinary surgeons judge the difficulty of a procedure, thus arriving at a contextualized understanding of skill. Brian Pentland describes the problem-solving strategies employed by technicians who staff software support hotlines. Stacia Zabusky investigates the difficulties microcomputer support technicians encounter as they attempt to construct and maintain a technical identity in the midst of an organization overwhelmingly populated by nontechnical users. Finally, Bonalyn Nelsen describes the situated moral understandings that guide emergency medical technicians through the uncertainties and risks they encounter in the course of their practice.

The vistas in Part II are less grand than those in Part I, but the resolution is much sharper. The reader will learn that the anomalies that surface in aggregate analyses of technical work in Part I are paralleled by, if not rooted in, disjunctures and ironies that pervade day-to-day life in technical settings. As we have already noted, no claim about technical work is more prevalent than the assertion that effective technical practice relies heavily on theoretical or abstract knowledge. Such claims are often used to justify requiring technicians to obtain degrees from postsecondary educational institutions in a relevant discipline. Pinch, Collins, and Carbone show that even in surgery, critical skills and judgments are contextually situated and formal knowledge is, at best, a small portion of what enables a surgeon to confront the ambiguities of practice successfully. Pentland further explores the disjuncture between formal and practical knowledge by explaining why support specialists have no choice but to approach software, the epitome of a closed rational system, as if it were contingent and poorly behaved.

Zabusky examines why technicians find it so difficult to garner respect for their skills and knowledge. She argues that microcomputer support specialists (and by implication other technicians who serve as brokers) inhabit a situation that breeds ambivalent identities. Microcomputer support specialists are simultaneously insiders and outsiders in the organizations in which they work. Their work places them in the ambiguous situation of being servants as well as experts. The pernicious dynamics of trust and distrust that surround such identities help us understand why technicians might appear in aggregate to be a cross between professionals and disgruntled employees. Yet, Zabusky hints at another dynamic that is largely obscured by aggregate analyses and that complicates any easy parallel between technicians and unskilled labor: the fact that technicians subscribe to a moral code that makes them incredibly conscientious, responsible, and even loyal employees. Nelsen's paper on EMTs complements Zabusky's analysis by explicitly examining how situated occupational ethics influence technical practice and the services that technicians provide. Nelsen shows that a shared moral code with clearly articulated notions of responsibility enables EMTs to navigate the disjunctures between autonomy and constraint and between expertise and servitude that characterize technicians' work. Moral understandings transform these disjunctures into situational dilemmas with a game-like structure of risk.

The papers in Part III pivot from examinations of practice to policy, demonstrating that close studies of technical work can lead to pragmatic as well as theoretical destinations. Mario Scarselletta's analysis of the nature of skill among medical technologists and technicians indicates how a situated appreciation of the disjuncture between formal and contextual knowledge may make for better regulatory and educational policy. Specifically, Scarselletta demonstrates why regulations designed to reduce laboratory errors by requiring more education of technicians and technologists are likely to be ineffective: they are predicated on a fundamental misunderstanding of the primary sources of error in laboratory work and on a misconception of what laboratory technicians actually do. Louis L. Bucciarelli and Sarah Kuhn develop a similar theme with respect to the disjuncture between engineering practice and education. They argue that the tendency to portray engineering as an applied science, when engineering is also a social process, has allowed engineering schools to become increasingly disconnected from the exigencies that confront young graduates. Leslie Perlow and Lotte Bailyn show that the disjuncture between the practice and ideology of engineering also adversely affects firms and the engineers they employ. Because "real engineering" is usually portrayed as an individualistic accomplishment focused on the world of objects, the contributions of engineers who facilitate the social dynamics that make engineering possible often go unrewarded. As a result, engineers who enjoy and are accomplished at the social aspects of engineering find themselves channeled into careers that are suboptimal for themselves and the companies for which they work.

The papers in Part III illustrate the problems that arise when decision makers attempt to regulate work and workers without considering the realities of the job. Because technical work, in particular, is so poorly understood, policy makers routinely fall back on stereotypes or images of work drawn from other occupations in order to make sense of technical work. When regulators mistake credentials for technical skill, when educators do not understand the work for which students are preparing, and when employers do not fully understand the practices and abilities necessary for technical work to proceed, they risk setting policies whose effects are not only unintended but sometimes so skewed that they exacerbate the problems they seek to resolve.

The potential cost of misunderstanding technical work extends beyond educational policies for specific occupations or the policies of particular firms. In recent years, influential and well-meaning economists and government officials have argued that the United States stands at the brink of a new era and must choose between a "high road" and a "low road" economy. The former is associated with a high skill/high wage labor force, the latter with a low skill/low wage labor force. The path to the "low road" presumably leads past increased reliance on foreign labor for manufacturing goods, further downsizings of the blue-collar and managerial labor force, and an emphasis on personal service industries. Traveling the "high road" is said to require embracing the notion that prosperity lies in developing a high-technology, information-based economy. The argument is that such an economy will demand highly skilled workers. Proponents of the "high road" scenario therefore argue for policies that will raise the skill level of the labor force, since higher incomes and standards of living should "naturally" follow.

The ethnographic evidence in this volume hints that such an approach may be less successful than its advocates suggest. First, between the conception and realization of a high skill/high wage economy falls the shadow of status. A number of chapters in this volume clearly indicate that skills are not necessarily recognized. As Zabusky notes, highly skilled technicians frequently find themselves cast in the role of servants. Even more troublesome, studies of technical work find only sporadic links between high skill and high wages. The salaries of some technicians lie well below that of blue-collar labor, even though no one disputes that technicians are better educated. And, as Keefe and Potosky illustrate, wage growth among technicians' occupations has not kept pace with increasing demand. In fact, in some technicians' occupations, wages have actually fallen since the early 1980s.

We suggest that the misapprehension of technical work and the potential implications of its growth is rooted in the continued use of categories and theories drawn from a now-fading industrial economy to discuss the nature of work in the emergent economy. By continuing to force new forms of work into a conceptual system appropriate for the declining blue-collar and clerical workforce, analysts reduce the odds that they will appreciate the way in which work, skills, and perhaps even economic dynamics have changed. Careful examination of practice suggests a need for new categories and a reconceptualized division of labor, at least for the industrialized nations. Specifically, we see a need to re-examine the supposition of tight links between wages, skill, and status. To the extent that emergent skills do not fit existing categories, they may not even be recognized as such, much less be rewarded. To assume that policy makers can facilitate a shift to a "high road" economy by simply facilitating the acquisition of skill and the emergence of high-technology, information-based industries seems naive. Technology, work, and skills apparently change more quickly than does our repertoire of concepts for understanding either the workplace, the labor process, or what people do for a living. In this final disjuncture looms the specter of policy failure.

# PART I

# Technical Work's Challenge to the Established Order

1

# TECHNICAL WORK IN THE DIVISION OF LABOR: STALKING THE WILY ANOMALY

*Peter Whalley and Stephen R. Barley*

## INTRODUCTION

As in all societies, Western notions of work rest on long-standing cultural distinctions, legacies of meaning which bind us to our past. Though we may not even be fully aware of them, distinctions between management and labor, profession and craft, blue- and white-collar, employee and self-employed, middle and working class, skilled and unskilled, permeate the way we talk, write, and think about work. They have utility precisely because we take them for granted. Such cultural categories shape our analyses, frame our data collection, and guide our policies and programs. They allow us to get on with the business at hand. From time to time, however, such cultural lenses may obscure more than they reveal, particularly when social contours shift in directions that differ markedly from those of the past. Under such conditions, relying on well-worn typifications may legitimate stances or courses of action out of step with the demands of the present. In this essay we argue that the rapid expansion and increasing economic centrality of technical work marks just such a time.

The difficulty of locating technical work within the boundaries of conventional notions of class has been discussed elsewhere (Smith 1987; Whalley

The work on technicians reported herein was supported under the Education Research and Development Center program agreement number R117Q00011-91, CFDA 84.117Q, as administered by the Office of Educational Research and Improvement, U.S. Department of Education. The findings and opinions expressed here do not reflect the position or policies of the Office of Educational Research and Improvement or the U.S. Department of Education. The work on engineers was supported in part by a grant from the International Division of the Ford Foundation (No. 74-05940) which likewise bears no responsibility for any of the findings or opinions expressed here.

1991a). We shall focus here on the difficulties faced in trying to comprehend technical work through two other cultural frames: (1) the categorization of work as either mental or manual which, in turn, undergirds the difference between blue- and white-collar work as well as the difference between professions and crafts, and (2) the notion that work must be either occupationally or organizationally structured. Technical work, we argue, transcends and destabilizes both dichotomies. Although societies have devised structures to manage the cultural anomalies that result, these resolutions have never been completely stable, particularly in the case of engineering. The difficulty of resolving the strains posed by technical work may become even more pronounced as the newly emerging category of technicians continues to expand.

## CULTURAL FRAMES OF REFERENCE

### *Mental and Manual*

One of the oldest stories we tell ourselves about work is that there are two kinds of labor: work done with the hands and work done with the head. Manual work, contaminated by its actual and symbolic association with dirt, material objects, and physical labor has long been accorded low status. Even when manual labor is intensely skilled, it has still been devalued because of its reliance on oral traditions and tactile understandings rather than more formal and abstract codifications. Mental work, by contrast, is clean and relatively privileged, largely because it involves working with symbolic representations of a "virtual" world abstracted and distanced from nature. When mental work is considered work at all—in preindustrial times people often viewed the manipulation of symbols as an accomplishment rather than a form of labor—it is seen as "civilized." Especially during periods when Western societies restricted access to the tools of symbolic manipulation, the ability to read, write, and do mathematics was both a source and a symbol of social status.

If manual labor has historically been the work of inferiors—slaves, peasants, and, more respectably, skilled craftsmen—mental labor has been the privilege of the elite: of gentlemen, and sometimes ladies, of aristocrats, officers, scholars, professionals and priests. It was disdain for manual work that condemned poor eighteenth-century French aristocrats to near starvation lest they should soil their hands with the plow. The same disdain encouraged nineteenth century English entrepreneurs to launder their wealth and enter into the leisured lifestyle of the country gentry (Wiener 1981). In the mid-twentieth century, disdain for manual work has shaped the education of elite engineers by minimizing their exposure to practical skills to prevent such skills from lowering engineering's status in the eyes of the academic and professional establishments (Calvert 1967; Ferguson

1992; Bucciarelli and Kuhn, Chapter 9). It has also played an important role in shaping the social and political allegiances of the clerical labor force (Kocka 1980; Mills 1956) and in fostering the sense of cultural inferiority felt by many blue-collar workers whose education provided little access to cultural resources (Sennett and Cobb 1972).

The superiority of mental or, perhaps better, "symbolic" work has never been simply a matter of prestige. From the beginning, writing and mathematics have been tools of social control. Writing made possible the development of the legal codes that allowed the rise of the early empires. The development of mathematics allowed the collection of levies and the keeping of accounts. These symbolic skills enabled the elite classes to "act at a distance" (Latour 1987) because they permitted abstraction from the specifics of the immediate context. Limiting access to symbolic skills further secured the power of the elite, as the medieval Catholic church explicitly acknowledged in its resistance to translating the Bible into the vernacular. Those who can use drawings, plans and other symbolic tools for controlling the manual workforce have often found places near the seats of power in industrial corporations.

### *Occupational and Organizational Images of Organizing*

A second cultural distinction that frames our understanding of work reflects the institutional context in which work is organized. Broadly speaking, we expect lines of work to be structured either occupationally or organizationally. The first term inscribes a world of crafts and professions; the second, a world of bureaucracies and firms. The former conjures up images of a relatively autonomous, collegial community of skilled practitioners; the latter, images of a hierarchy of authority and jobs of indeterminate skill linked together by well-structured career ladders.

In an occupationally ordered world, knowledge and skills are assumed to be specific to particular domains and too complex for nesting in a single hierarchy. As a result, authority and expertise are treated as the property of distinct occupational groups (Freidson 1973a). Because individuals, rather than positions, serve as vessels of expertise (Abbott 1991), knowledge is transmitted through extended apprenticeships or formal schooling in specialized curricula. Coordination occurs not through a chain of command, but through the collaboration of members of different occupations working jointly, as in the case of the modern building trades (Stinchcombe 1959).

By contrast, in an organizational division of labor, authority and expertise are arranged hierarchically. People in higher echelons not only have power over those below but, at least in an ideal bureaucracy, they legitimately exercise authority only to the degree that their knowledge is seen to encompass and exceed that of their subordinates (Weber 1968 [1922]). Organizations, rather than members of an occupation, become the primary means for preserving expertise because knowledge is encoded in rules, roles and procedures which are invested in positions, rather than people. Because such knowledge is often highly specific and valuable to an organization, internal labor markets evolve to define careers and ensure loyalty (Doeringer and Piore 1971). This means of organizing work epitomized the Prussian civil service on which Max Weber modeled his theory of bureaucracy, and informed Frederick Taylor's attempts to manage the modern factory "scientifically." It is also the model of work that once made "I work for IBM," a more acceptable answer to the question, "What do you do?" than the response, "I am an engineer" (Whalley 1987).

In the preindustrial world most work was organized occupationally. Even agricultural expertise was divided according the crops that farmers grew (Applebaum 1992). The occupational system was best developed, however, among the skilled crafts whose domains were secured by guilds or guild-like collectives (Sonenscher 1987). The occupational communities that the guilds regulated were often marked by considerable specialization and an extensive apprenticeship system. Although the division of labor in preindustrial society had vertical overtones, these typically reflected either the relationship between master and apprentice or the workings of an omnipresent class structure, as in the case of serfs and lords. The organizational structuring of work was, by comparison, weakly developed. Only in the Roman Catholic church and certain European militaries was work structured along organizational lines. The modern world, by contrast, is a world of organizations, state bureaucracies, and multinational corporations. Indeed, until quite recently, most social theorists viewed the economic and political development of the twentieth century largely as a story of the ever-increasing dominance of large organizations (Chandler 1977; Schumpeter 1942; Williamson 1975).

Organizations and occupations, however, are more than alternative ways of dividing labor. The contrast between them also shapes our images of work. In the United States and Great Britain, for example, the ideal of professions and crafts being the natural and desirable way to organize work has continued to exert considerable influence, even as organizational forms have come to predominate. The governments of both countries continue to collect statistics on the labor force in occupational terms (Udy 1981; Whalley 1987), and members of both cultures tend to associate autonomy, freedom, and financial independence with the occupationally structured work of the crafts and professions. In Anglo-American society, few insults are worse than calling someone a bureaucrat, with the term's connotations of colorless individuals who have sacrificed their souls to an organization. Organization men are frequently portrayed as people who are closely monitored and trapped by routines, who carry out orders without question, and wait only for retirement.[1] In continental Europe and Japan, by contrast, the image of the civil servant, the quintessential bureaucrat, has proven far more attractive. There, bureaucracy is associated with the noble traditions of the military, the church, and the building of the state. The French "cadre" (a social category archetypically composed of middle-level bureaucrats), enjoys the same social prestige as do professions in U.S. and British society (Boltanski 1987).

These distinctions—between mental and manual, and organizationally and occupationally structured work—are cultural frames of great power. They affect the way we see, think about, and value the work we do, and have important social and practical consequences for both private and public policy. The cultural gulf between mental and manual work, for example, has led to the gradual degradation of any work that requires practical skills, and to the subsequent overvaluation of white-collar jobs. As a result, educational programs have bifurcated into college-preparatory and vocational tracks: the first prestigious but entirely symbolic in content, the latter looked down upon and underfunded because of its focus on practical skills. If an increasing number of jobs come to meld mental and manual work, however, such a division may lead to a disjuncture between the society's educational policy and the requirements of its labor market. Similarly, if work can no longer be separated into jobs that are occupationally or organizationally structured, then individuals' career expectations, firms' employment policies, and society's responsibilities for education and training could all become seriously distorted.

Our thesis is that the expansion of technical work poses just such a challenge to these cultural dichotomies, both practically and analytically. We shall argue that this challenge has occurred in two waves. The first began with engineering's emergence as a distinct industrial activity in the latter part of the nineteenth century. The second arose with the appearance of technicians as an identifiable occupational category after World War II. Engineers, we argue, may ultimately have found a reasonably secure but ambivalent identity as a new kind of corporate professional—although that identity, too, looks increasingly unstable. Technicians, however, continue to pose major practical and intellectual challenges to our images of work.

## THE CULTURAL CHALLENGE OF ENGINEERING

Modern engineering is the child of industrial capitalism. Although commentators sometimes point to its continuities with civil and military engineering, which have much longer histories, engineering as we presently know it emerged

1. The presence of women in organizational careers is still sufficiently recent to carry the glamour of pioneering, so perhaps this cultural stereotype is less applicable to them.

alongside the corporation in a world where work was still organized occupationally. In fact, engineering played an important if ambivalent role in the genesis of industrially organized labor.

Although the guild system had largely disappeared by the early nineteenth century (and much earlier in England), many of its traditions lingered at the dawn of industrialization, especially among the skilled trades and learned professions. Until late in the nineteenth century control over production largely remained the preserve of craftsmen (Nelson 1975; Clawson 1980), who passed on their knowledge and skill, which was almost exclusively tacit and oral, by training apprentices (Sturt 1923). In the early days, craftsmen were often themselves employers. Those who were not often contracted with employers for control over portions of the production process, giving them authority over the work as well as unskilled laborers and apprentices.

With the onset of industrialization, an organizational order slowly began to supplant this occupational order. Firms sought to rationalize the tacit, oral, and secret skills of the crafts and to locate craftsmen within a vertical, rather than horizontal, division of labor. Semi-skilled workers, whose activities were regulated by machines and engineers, gradually replaced those craftspeople who had stood at the heart of the production system: ironworkers, tailors, seamstresses, machinists and others (Edwards 1979).[2] The process involved tearing apart older, more integrated craft techniques so that design and planning could be separated from actual manufacture. It also entailed developing new and more symbolic methods that permitted the control of manufacture from afar. This separation of planning from execution led to control over the labor process being vested in managers whose technical expertise was sometimes more theoretical than practical. It also led to the establishment of drawing offices where machines and production processes were often designed and developed with the goal of tighter regulation of the shop floor (Ferguson 1992).

Such a radical reorganization of work required the creation of a technical staff knowledgeable in design and production techniques, committed to the employers' interest, and willing to assist management in controlling the industrial workforce. Technical staff began to control both the speed and the methods of production, a process carried to its logical, if sometimes irrational, conclusion by the practitioners of scientific management (Braverman 1974). Later, research and development laboratories began to institutionalize the development of new products by incorporating into the organizational division of labor the work of the inde-

2. A few crafts successfully retained their privileges as a labor aristocracy, in part because they were more willing to cooperate with employers. Placed in charge of the maintenance of factories but located at the margins of the production process, electricians, pipe fitters, and other tradespeople specializing in building and repairing equipment, retained traditions of apprenticeship and distinct occupational identities.

pendent inventor, who, like the craftsman, had often integrated design and practice (Reich 1985, Whalley 1991b). Thus, the emergence of engineering, and the development of the industrial corporation, with its organizational division of labor, are simply two sides of the same historical process.

Although modern engineering emerged hand in hand with industrialization, a serious history of the way in which modern-day engineering's various functions came to be bundled together remains to be written.[3] Certainly the military engineers of the Renaissance provided one model for the integration of design and command. Especially in France, schools of military engineering provided an important prototype: first for the civil engineers who built the roads, canals and railroads of the nineteenth century and, later and less directly, for mechanical engineering (Weiss 1982). In the United States and Great Britain, where formal education in engineering developed later (Calvert 1967; Smith and Whalley 1996), the artisans who built the early machines of the Industrial Revolution were more important role models. These early mechanics were not members of a formally educated officer class, trained in math and drawing, but practical tradesmen who only reluctantly supplemented their empirical approach, rooted in tacit and oral knowledge, with the skills of formal drawing needed for mass production. Although the story is sometimes written as if these mechanics were simply deskilled and replaced, many were absorbed into the new technical positions. They were supplemented, especially in the United States, by the relatives of entrepreneurs who saw a professional career in industry as the modern equivalent of the older aristocratic career in the military. All that these groups had in common was an interest in designing and planning and a willingness to apply that interest to the corporate organization of production.

Out of this mélange, engineering grew to become one of the largest and best-established professions of the twentieth century. Like other professions, engineering now boasts its own formal training programs in universities and its own professional societies. Engineers are relatively well paid and enjoy a degree of job security as well as considerable social respectability. Furthermore, engineering is firmly situated on the management side of the organizational divide that has traditionally separated white-collar from blue-collar, "staff" from "works," "suits" from "working stiffs" and "exempts" from "non-exempts."[4] In short, en-

3. Such a history would require the kind of approach recently developed in constructivist accounts of the emergence of science disciplines (Latour 1987; Porter 1977), rather than one that prejudges the way in which various tasks were bundled into present-day occupations, or assumes that engineering was a potentiality waiting to be actualized by technological and social developments. The latter, unfortunately, has been the style of many histories of the profession.

4. Engineering has been institutionalized most fully as a "corporate profession" in the United States. In Britain, the intertwining of engineers with positions that would be filled by technicians in the U.S. has left many engineers with a weaker claim to professional status. British engineers are better understood as part of a larger corporate staff which encompasses a wider range of nonmanual

gineers have become what Peter Whalley (1986b) has called "trusted workers." They are collectively, though not necessarily individually, loyal to their company and share, in good times at least, many—though not all—of the political interests of senior management. In fact, engineering is something of a model for a range of newer corporate professions such as accounting, marketing, and personnel management.

At first glance, then, engineering seems to fit our conventional cultural categories for classifying work rather well. White-collar and even professional, engineering is so tightly linked to science—often thought of as the archetypical "mental" work—that the public often finds it difficult to separate the two. Engineers have also been so central to industrial organizations that many firms considered to be icons of the American economy—IBM and General Motors, for example—are widely perceived to be embodiments of engineering culture. Yet, if we look more closely at engineering, several discrepancies emerge that are rooted in the peculiar attributes of technical work.

### *Anomalies*

*Mental or Manual.* Although engineering may be firmly ensconced as a twentieth century profession and the curricula of engineering schools may center on the learning and application of scientific principles, the occupation has never fully escaped its manual legacy. Engineering has never been quite abstract or symbolic enough to be a purely "learned" profession. Unlike lawyers, scientists and clergy, engineers have always been involved in practical matters, in making *things* and making them work. If mathematical or scientific analysis assists the engineer, all the better. But if successful practice is shown to depend on trial and error or on local and contextual knowledge, then that too has generally been acceptable to most engineers (Ferguson 1992; Henderson 1991; Vincenti 1984). Engineering is technical work not simply because engineers rely on esoteric techniques and instruments, but because machines and systems are engineering's ultimate product, even if, in a siliconized world, some machines have a virtual rather than a mechanical reality. Making existing machines work, designing new ones, and doing these tasks well, remains at the heart of engineering culture (Kidder 1981; Kunda 1992).

---

work. Nonetheless, many of the distinctive social arrangements that characterize corporate professions in the United States are also in place in Britain. These practices serve to secure corporate loyalty and separate technical staff from manual workers (Whalley 1991a). In France, engineers have an even higher status than they do in the States. As in the U.S., the French notion of cadre excludes technicians but includes a broader range of management positions and stresses organizational authority rather than technical expertise (Boltanski 1987; Crawford 1989).

This commitment to machines and practical activity has always made it difficult for engineering to depend solely on college education for training. In all but the most advanced areas of high technology, practical experience continues to play an important role. Engineers retain a grudging admiration for the self-trained technical wizard who knows almost intuitively how to design and build machines and get them to work. Studies in a variety of industrial settings and in countries with different educational systems have consistently found that about a third of all practicing engineers are former craftspeople and technicians with no formal training in engineering (Crawford 1989; Whalley 1986b; Zussman 1985). Although this percentage may be declining as older engineers retire and as pressures for credentials increase (Keefe and Potosky, Chapter 2), much credentialling still occurs *after* the individual has entered the occupation.

This "manual" side of engineering has always been most visible in Britain. Early in the nineteenth century the Mechanics Institutes sought to integrate practice and theory by exposing artisans to the world of equations, and university-trained engineers to the world of manufacturing (Wrigley 1986). By the end of the century, however, the middle classes, wary of the status implications of working too close to machines, had distanced themselves from engineering (Ahlstrom 1982; Donnelly 1991; Wiener 1981). As a result, university-trained engineers played only a minor role in industrializing Britain. The lack of university-trained engineers, in turn, encouraged British firms to rely more heavily than elsewhere on apprenticeships and promotion from within (Smith 1987; Smith and Whalley 1996; Whalley 1986b). Indeed, employers often resisted efforts to formalize engineering education, on the grounds that it would distance engineers from industrial practice. British engineers still retain strong links to craft traditions and identify more closely with the practical aspects of their work than do engineers in other countries (Whalley and Crawford 1984). Many are even members of unions aligned in some fashion with the crafts (Smith 1987).

In the U.S. status barriers were less strong than in Britain and the manual attributes of engineering were initially less stigmatizing. The children of small businessmen entered engineering in larger numbers and the owners of firms were often proud of their hands on know-how (Noble 1977). The social acceptability of mechanical acumen allowed engineers to acquire social status within a "shop culture" more typically present in the crafts (Calvert 1967). Only the rise of the professional school—located in universities but accredited by professional bodies with strong employer representation—transformed engineering's culture of practical experience into one of formal education (Layton 1971). The transformation was not without its critics. Many have argued that the search for academic respectability—the conversion of engineering from a practical art to a symbol-manipulating "science"—has weakened the engineer's skills and distanced academic engineering from the employer's needs (Ferguson 1992; Bucciarelli and

Kuhn, Chapter 9).[5] As did their British counterparts, American engineering educators attempting to promote the value of a more abstract and scientific education have run into strong resistance from both employers and many practicing engineers.

The practical nature of many technical tasks has not been the only source of continued tension between the professional and craft aspects of engineering. Marxist analysts, for example, have pointed to engineers' dependent status as employees, their location in staff rather than supervisory positions, their craft-like apprenticeships, and their occasional propensity to unionize, leading some writers to portray them as members of a "working class." For reasons suggested above, the claim has been most plausible for British engineers (Smith 1987), but elements critical to the working-class argument are present to some degree in all countries. For example, upon observing the radical moment that resulted in the French student revolt of 1968, Serge Mallet (1975) and André Gorz (1967) pronounced technical workers part of a "new working class" whose knowledge and self-confidence were sufficiently intact for them to resist incorporation into capitalism.[6]

Trying to ignore distinctions between engineers and manual workers, however, has proved as difficult as treating them as purely "mental" employees. Despite their practical bent, engineers value the ways their "book learning" and mathematical abilities contribute to their work and status (Robinson and McIlwee 1991). Engineers jealously guard their privileged position in the workplace. They have rarely been willing to unite with others, especially manual workers, in order to bargain collectively. Moreover, the sporadic infatuation during the 1960s with employee participation among engineers often proved to be a strategy for improving their own status in managerial affairs rather than a plea for worker solidarity (Crawford 1989, Whalley 1986a).

*Occupational or Organizational Work.* Attempting to portray engineering in solely occupational or organizational terms is no less difficult than the Procrustean exercise of forcing it to fit traditional notions of mental and manual work. In the United States and Britain, at least, engineering is usually cast as a profession. Proponents of this view point to such occupational trappings as cre-

5. This is not just a practical problem for educators. The "craft" of engineering has long posed difficulties for analysts who have sought to place engineering among those professions that owe their status and power to scientific knowledge (Freidson 1973a; Larson 1977). Indeed, the very attempt to use "science" as a way of distinguishing professions and crafts runs afoul of ethnographic evidence that even the "pure sciences" have their practical and instrumental, perhaps one should say "technical," side. If even the boundaries of science are socially constructed, it is hard to seek the essence of professionalism this way (Whalley 1991a).

6. Erik Wright (1985), writing in the American context, also saw engineering as a "new" craft, but was less sure of its place in a unified working class. He saw engineers, like the old labor aristocracy, using their skills to "exploit" other workers and obtain more than their fair share of the surplus.

dentials, professional schools and associations, and engineering's distinct knowledge base. Historically, many engineers have indeed sought to work as technical experts unencumbered by administrative and managerial responsibilities. These engineers are often more deeply committed to their technical communities than to the organizations that employ them. There is some evidence that an occupational orientation may even be increasing among engineers, especially in high-tech industry (McIlwee and Robinson 1992). Such engineers read technical journals, get excited by technical advances, and can be recruited with offers of opportunities to "play" with the latest technology. Especially early in their careers, many engineers express disdain for the administrative component of management. There have even been historical moments when engineering's organizational status has been placed in doubt by professionalizing movements (Layton 1971; Meiksins 1989). For example, despite its seeming corporate agenda, even scientific management can be interpreted as an attempt to assert engineers' unique occupational identity and interests (Stabile 1984).

Paying too close attention to engineering's occupational and technical culture, on the other hand, can lead one to overlook engineering's organizational character. Even design, the most visible and arguably the most professional of all engineering functions, is deeply embedded in an organizational context. Design emerged as a distinct task only with the disaggregation of craftwork. The engineering drawing was itself developed as an organizing tool (Ferguson 1992), and continues to serve that role in the structuring of modern production systems (Henderson 1991). Furthermore, engineers have never been restricted to design. Mass production also involved the detailed planning of production processes—the function championed by Frederick Taylor (1911) and other advocates of "scientific management"—which gave industrial engineering, in particular, a distinct managerial aura. In France, where the state sponsored industrialization and the French Revolution banned most craft apprenticeships, engineering clung to its preindustrial origins in the French officer corps. Their training in prestigious technical schools places French engineers at the core of the managerial class (Weiss 1982). Even in Britain, university-trained engineers are increasingly involved in the management of production (Whalley 1986a). Finally, there is much evidence that even design engineers readily show interest in management when technical and financial rewards encourage them to climb the organizational ladder (Whalley 1990).

The engineer's tendency to combine occupational and organizational interests has frustrated sociologists on both sides of the Atlantic. As a generation of American sociologists discovered, attempts to treat engineering as a traditional profession founder on the engineers' lack of commitment to a set of "professional" values that might put them at odds with their employers (Perrucci 1969; Ritti 1971). Unlike other professionals whose occupations establish their identities outside of corporate employment (scientists and physicians, for example), engineers seem to be remarkably contented employees. They show no particular desire to share their knowledge with the public at the expense of the corporation, find little wrong with the profit motive, and are willing to pursue the rewards of a managerial career even when it draws them reluctantly away from technical work. If engineers are professionals, they are committed "corporate professionals" rather than members of a traditional "free" profession.

Among French sociologists it has been engineering's occupational propensities that have caused intellectual problems. French engineers have traditionally been members of an elite cadre, firmly embedded in the higher ranks of state and corporate hierarchies. What Mallet and Gorz glimpsed in the 1960s, however, was the emergence of new groups of French engineers less firmly embedded in the bureaucracy. Unlike traditional French engineers who managed production facilities, the new engineers (who often worked in R&D labs or aerospace manufacturing) served as technical experts in staff positions. For many of them, being in the position of staff felt like a loss of privilege in the formal hierarchies of French corporations. The historical moment when many French technical staff joined in the strikes and demonstrations of May 1968 was short-lived, but it forced the intellectual recognition that engineering encompassed functions not adequately characterized as "management" with the term's intimations of a close alliance with employers. If neither a "new working class" nor members of the "old" middle class, post-sixties structural Marxists portrayed them as a "new middle class," occupying a "contradictory class location" (Poulantzas 1975; Carchedi 1977). If engineers are partly management, they are also partly workers and, perhaps more importantly, workers who possess a distinctive kind of occupational culture.

Engineers thus continue to play havoc with conventional models of work even though they are the prototypical technical workers of high industrialization. Engineering is too practical and machine-oriented to be a profession of the text, yet, too numerate and abstract to be dismissed as "mere" manual work. Engineers are too organizationally comfortable and career-oriented for their work to be fully understood in occupational terms; yet, engineers are too committed to their own technical culture to be fully assimilated into the blue-suited ranks of management. Thus, even though engineers may be corporate professionals, the underlying difficulties posed for our conventional cultural frames by the technical nature of their work have not been entirely resolved.

### *The Future of Engineering*

Ultimately, the formation of engineering as a corporate profession needs to be understood as a social construction that finessed opposing cultural images of the nature of work without fully resolving the tensions embedded in them. Even so, construing engineering in this way exacted a heavy price from corporations. Firms only retained engineers by granting them costly pay and benefit packages, promising relative job security, and creating opportunities for engineers to ascend into management. Historically, employers had a number of reasons for adopting this strategy. First, because engineering played a key role in the struggle for control of the shop floor, employers needed the engineers' loyalty. Second, because of the explosive growth of the economy during most of the twentieth century, firms could easily create enough middle-management positions to offer engineers promotional opportunities. Finally, even though engineering was the largest of the corporate professions, in all but a few high-tech companies engineers have represented only a small part of a firm's labor costs.

Questions remain, however, as to what might occur if the social contract between engineers and corporations were to unravel, or if new categories of technical workers were to arise for whom managerial careers were inappropriate or too expensive. The first question seems ever more germane given the recent downsizing and reorganization of American corporations. Over the last decade, implicit promises of job security have been abandoned, middle-management positions have been decimated and firms appear ever more willing to treat engineers as temporary technical contractors. This may well have a major, but as yet unresearched, impact on the structure and culture of engineering.

The second question raises the issue of the social location of that newly emergent group of technical workers: technicians.

## THE ADVENT OF TECHNICIANS' WORK

Although a handful of technicians' occupations existed at turn of the twentieth century, most lines of work thought of as technicians' work came into being over the last forty years.[7] In fact, the modern meaning of "technician"—a person whose work revolves around instruments and who requires specialized training in a science or technology—did not appear in English-language dictionaries until after World War II (Barley and Orr, Introduction). In 1950, technicians comprised 1 percent of all employed Americans (Szafran 1992). By 1990, the percentage had grown to 3.4 percent. The U.S. Department of Labor estimates that the percentage will rise to nearly 4 percent by the middle of the next decade (Silvestri and Lukasiewicz 1991). These figures indicate that the proportion of Americans employed as technicians has grown by 240 percent since mid-century, a rate that dwarfs the expansion of all other occupational clusters charted by

7. For example, radiological technicians were recognized as members of a distinct occupation between World War I and World War II (Larkin 1983). Shapin (1989) has shown that as early as the seventeenth century, noted scientists employed individuals in their labs who filled roles roughly analogous to those of today's science technicians.

the Bureau of Labor Statistics. By way of comparison, service and professional occupations, which rank second and third for growth over the same time period, expanded by 89 percent and 82 percent respectively.[8] To be sure, technicians still constitute one of the smallest occupational categories, but the number of technicians can hardly be considered insignificant. Technicians now outnumber farmers in the American economy, and in some industries technicians are surprisingly numerous. For instance, as of 1990, American industry employed one engineering technician for every two engineers, one science technician for every two scientists, and two health care technicians for every physician.[9] In medical settings, technicians and technologists even outnumber registered nurses—yet we know less about technicians than about the members of any other occupation.

### *Social Origins*

Like engineers, technicians usually work in organizations dominated by members of other occupations. However, because "technician" is a much broader occupational classification than "engineer," technicians' work is more diverse and less easily characterized than engineering. The paths by which technicians' occupations have evolved are also more numerous. Nevertheless, most technicians' occupations appear to have arisen by one of six analytically (but not necessarily empirically) distinct paths.

*Hiving off of Work.* Medicine, law, engineering, science and most other widely recognized professions have witnessed a tremendous expansion and specialization of knowledge over the last forty years. In response, overburdened professionals have sought to lessen their workload, in part by allocating more routine tasks to members of other occupations. Many of the technicians' occupations that have flourished in the latter half of the twentieth century originated in the hiving off of work by the established professions.[10] The phenomenon has been most visible in health care, where licensed practical nurses, medical technologists, sonographers and an expanding array of other technicians have coalesced into occupations around tasks discarded by physicians and registered nurses. The dynamic is also prevalent outside health care, where it has given birth to a plethora of occupations ranging from the well-known (paralegals, computer programmers) to the amazingly obscure (test and pay technicians; see Kurtz and Walker 1975).

8. We have calculated these growth rates using data on an occupation's share of the labor force found in Szafran (1992) and Silvestri and Lukasiewicz (1991).

9. These ratios were estimated using detailed occupational census data for 1990 published in Table 2 of Silvestri and Lukasiewicz (1991).

10. The term "hiving off" is adopted from Smith (1987). The concept, though not the term, entered the sociological literature with Hughes (1958).

*De Novo Creation.* Other technicians' occupations have been created *de novo* in the sense that their work was not previously performed by members of an existing occupation. Almost all instances of *de novo* creation are rooted in the invention and subsequent commercialization of a technology. Prior to the invention of photocopying, there were no copy repair technicians (Orr 1991b, 1996). Prior to the diffusion of microcomputers, there was no need for microcomputer support specialists. The spread of television similarly stimulated a market for television repair. Although many technicians' occupations that have arisen *de novo* center on the maintenance of a technology, not all do. For instance, sonographers rarely repair ultrasound equipment, yet their work arose with the use of ultrasound in medical imaging (Barley 1990). Air traffic controllers, EEG technologists, and technicians who monitor the controls of nuclear power plants are further examples of occupations, spawned by new technologies, that have little role in the technology's maintenance.

*Occupationalization of Amateur Work.* A third group of technicians owe their positions to the occupationalization of amateur work. Radiological technologists are a case in point. At the turn of the century, anyone who could afford a cathode ray tube could establish a business providing x-rays for medical diagnosis (Larkin 1983). Not only did numerous engineers and physicists set up shop as purveyors of medical images, but after World War I medics trained to operate x-ray equipment on the battlefields of Europe returned home to establish medical imaging practices. Gerald Larkin (1983) reports that in the years following World War I, it was fashionable for people in European and North American cities to frequent such shops to secure x-rays as curiosities. The occupation of the radiological technologist arose, in part, as physicians lobbied for laws that drove amateurs and free lancers from the market (Barley 1986).

Emergency medical technicians are a more recent example of a technical occupation that evolved from amateur work (Metz 1981). Before the 1970s no EMTs existed. Emergency rescue services outside of large cities were almost always staffed by volunteers charged with transporting the sick and injured rather than with performing triage. At best, members of rescue squads were trained in first aid. In response to a public outcry over the increasing number of highway fatalities, during the 1970s the federal government urged the states to create a corps of trained emergency personnel modeled after the military paramedics who had proven themselves capable of dramatically reducing mortality rates in Vietnam. All states now train and certify emergency medical technicians. Although many EMTs remain volunteers, for a variety of reasons the work is rapidly shifting from volunteer to paid personnel for whom emergency medicine has become a vocation (Nelsen and Barley 1993).

*Upgrading of Mechanics.* The upgrading of mechanics and other craftspeople constitutes a fourth process by which technical occupations have formed. Steven

Shapin (1989) has shown that scientists employed laboratory assistants as early as the seventeenth century. Until recently, however, laboratory assistants were usually mechanics who specialized in the fabrication of instruments and other experimental apparatuses. In fact, many laboratories continue to employ machinists. The role of the modern science technician emerged as lab mechanics assumed responsibility for the design and execution of experiments. Separation of the two occupations solidified in the early twentieth century when it became routine for organizations to require technicians to have an advanced degree in a relevant discipline (see Keefe and Potosky, Chapter 2). With the occupation established, a hiving off process has more recently driven the expansion of the science technicians' duties and skills. A similar phenomenon surrounds the birth of engineering technicians and draftsmen. Peter Whalley (1986b) notes that, even today, firms often draw engineering technicians from the ranks of machinists who have demonstrated analytic talent.

*Technization of Work.* A fifth and more recent impetus for the formation of technicians' occupations has been the technization of work—the transformation of traditional blue- and white-collar work by microelectronic technologies (Barley 1993). Accumulating evidence indicates that complex information systems infuse nontechnical and even semi-skilled work with technical content. Shoshana Zuboff (1988) concluded from her studies of paper mills that computer-integrated production requires blue-collar operators to analyze data and make decisions based on their analyses in order to control production processes. In the past, such skills were reserved for middle managers. Analogous findings are common to most other studies of manufacturing plants that have adopted computer-integrated controls (Hirschhorn 1984; Majchrzak 1988; Kern and Schumman 1992). The central dynamic in such settings appears to be the need for operators to rely on digitally encoded symbols rather than sensory data for monitoring and managing production systems.

*Regrading for Social Control.* Finally, issues of social control have stimulated the creation of some technicians' positions. A growing number of firms have begun to reclassify the work of factory operatives. The regrading of work in the absence of technical change entails little more than awarding operatives a new title, sometimes in the hope of avoiding unionization. The origins of Proctor and Gamble's practice of calling its factory operatives "technicians" is a case in point.[11] During the early 1970s, a period of relative labor unrest, Proctor and

11. The following information was contained in a lecture given to a group of academics (including the second author) attending a three-day workshop on Total Quality Management practices at Proctor and Gamble, June 1–3 1993. Many of Proctor and Gamble's production lines now depend on advanced microelectronic controls. As a result, technicians' jobs appear to have been transformed through the technization of work and may now warrant the name they were originally given for other purposes.

Gamble decided to build a number of new plants in right-to-work states in the South. The firm formally titled the operators who staffed these plants "technicians." Although the southern "green field" sites were equipped with highly automated technologies, management's decision to call operators "technicians" was unrelated to the requirements of their jobs. In fact, the firm did not reclassify operators in unionized plants in the north who worked with similar technologies. The firm dubbed southern workers "technicians" as part of a plan to forestall unionization since technicians are typically counted as "exempt" employees who fall outside the purview of U.S. labor law.

### *Diversity and Commonality in the Social Organization of Technicians' Work*

Since no collective can escape the constraints of its past, we would expect the diverse origins of technicians' work to influence how technicians are trained, the structure of their careers, the probability that they will unionize, their social identities, and so forth. For instance, all else being equal, technicians whose work has been hived off from a profession should be most inclined to emulate professional forms of organizing. In fact, most of the technicians' occupations that use credentials as barriers to entry and have their own journals, training programs, and occupational associations are found in medicine, where hiving off has been most common. In contrast, occupations created *de novo* by a new technology should have relatively few barriers to entry and little formal structure. Occupations that focus on the repair of new technologies are especially likely to be open to anyone with mechanical acumen. For this reason, one might expect many people who repair computers to be self-trained. When the technology that occasions the *de novo* creation of a technical occupation becomes sufficiently widespread, occupational identities and fledgling communities of practice may evolve (Orr 1991b). On average, however, such occupations should evince fewer trappings of a profession than occupations created by a hiving off process.

Compared to other technicians, those whose work has been altered by a process of technization or who perform duties formerly performed by amateurs should have the greatest difficulty establishing a recognized occupational identity, albeit for different reasons. Lines of work that have been infused with technical content by microelectronics generally have previously existing identities and statuses. Because those who do not do the work are unlikely to appreciate how the work has changed, perceptions of the technicians are likely to be constrained by existing cultural frameworks. Zuboff (1988) reports precisely such a dynamic in her study of pulp paper mills where many managers were unable to admit that the operators' work had been radically transformed by computer-integrated manufacturing. These managers were uncomfortable granting operators greater responsibility and the operators themselves were hesitant to shed well-understood blue-collar identities and attitudes. Occupations arising from amateur work may also have more trouble acquiring a skilled identity than occupations with different origins, because they face publics who are not only unaccustomed to distinguishing between amateurs and full-timers, but who may be hesitant to attribute skill to an activity formerly performed without pay. (Nelsen and Barley 1993). Consequently, conscious attempts to construct skilled identities should be most common among technicians involved in occupationalizing amateur work.

Finally, one would expect workers who become technicians by regrading alone to exhibit relatively traditional blue-collar and lower white-collar identities, since a name change is all that has occurred. Among such technicians, employment relations should be dominated by bureaucratic control, and unionization should be most common. Career lines, if they exist at all, should terminate at first-level supervision. Training should be confined to on-the-job learning and should not differ significantly from that typical of most blue-collar or clerical work. Exceptions to this pattern should occur only when employers couple the regrading of work with an attempt to redesign work systems and employment practices.

Although technicians' occupations vary with respect to origin and social organization, most still share a number of critical characteristics (Barley 1992). With the exception of factory operatives whose jobs have been regraded, few technicians are directly involved in the production of goods. At most, technicians repair or monitor the technologies that undergird production systems. Since few people who have not worked as technicians know what technicians know, their work generally appears esoteric to outsiders. Hence, technicians are usually recognized as skilled individuals. But most important, individuals employed as technicians almost always work on, with, or through reputedly complex technologies and techniques to manage an interface between a larger work process and the materials on which the process depends. Technicians traffic in symbols, data and diagnoses. They assist professionals, vend information, provide consumers with sophisticated services, and maintain complex technologies. Yet, unlike the popular image of a knowledge worker whose work is entirely symbolic, technicians also remain intimately involved with the material world. Technicians work routinely with machines, human bodies, and a host of other physical systems. Their encounters with the material world are as likely to be physical as they are to be mediated by instruments, algorithms, and techniques. Thus, technicians violate the same cultural categories that engineering challenged over a century ago.

### *The Anomalies of Technicians' Work*

*Mental versus Manual.* Like engineering, technicians' work blurs the boundary between mental and manual work and, by extension, the dividing line between blue- and white-collar labor. Technicians often wear white collars, carry briefcases, and conduct sophisticated scientific and mathematical analyses. Yet they use tools and instruments, work with their hands, create objects, repair equipment and, from time to time, get dirty. Like those in positions of authority in organizations, technicians have considerable autonomy and are often trusted by their employers (Barley and Bechky 1994). Yet, like those in lower echelons, technical workers are often poorly paid (Franke and Sobel 1970), are accorded low status (Zabusky and Barley 1996), and may be subject to stringent controls (Orr 1991b).

As in engineering, the technician's melding of mental and manual work also blurs the distinction between craft and profession. Table 1.1 describes this blurring by situating technicians' work with respect to characteristics that sociologists traditionally attribute to crafts and professions. Technicians resemble professionals in that their work is esoteric enough that few outsiders can claim to possess their skills or knowledge. It is also relatively analytic and often requires specialized education. In fact, with the exception of professionals, technicians are the best-educated of all occupational groups (Carey and Eck 1984). The average technician in the United States now possesses two years of post-secondary education. Like the professions, some technicians' occupations have developed occupational societies and journals.

Yet, in other ways, technicians' work more closely resembles a craft. Apprenticeships and on-the-job training play a crucial role in the education of technicians, just as they do in the training of craftspeople. In fact, many technicians are trained solely through apprenticeship. For instance, Stacia Zabusky and Stephen Barley (1996) found that over half of the microcomputer technicians they studied had no formal training in computers or electronics. Like craftspeople, technicians operate equipment, create artifacts, and possess valued manual skills. Technicians also resemble craftspeople in that they are more likely to unionize than are most other nonclerical white-collar employees, a tendency that is especially strong in Europe.

*Organizational versus Occupational Alignment.* Like engineering, technicians' work challenges the traditional dichotomy between vertically and horizontally organized labor. Most technicians' occupations originated in an organizational context and most technicians continue to be employed by firms, hospitals and other vertically structured organizations. It is impossible to estimate with existing data the number of technicians who work for organizations populated entirely

TABLE 1.1

Characteristics of professions, crafts, and technical occupations

| *Attribute* | *Professions* | *Crafts* | *Technical Occupations* |
|---|---|---|---|
| Skills and knowledge possessed by people outside occupation | Knowledge and skills are esoteric and well guarded. Few outside the occupation have more than a trivial understanding of the content of the occupation's knowledge base. | Basic skills and knowledge are widely held by persons outside the occupation. However, finesse is less widely distributed. | Knowledge and skills are esoteric. In some instances, amateurs may exist, but in general, they are relatively rare. |
| Balance of mental/analytic versus manual/sensate work | Tasks are heavily weighted toward mental and analytic work. | Tasks are heavily weighted toward manual and sensate skills. | Tasks involve a heavy mental and analytic component, but the work also often has a signficant manual or sensate component. |
| Importance of formal education as a means of training and socialization | Most require either specialized undergraduate or graduate training. All require a college degree. | May require a formal apprenticeship. Otherwise, formal education is irrelevant. | Most require either a bachelor's degree or a specialized associate's degree or its equivalent. |
| Importance of on-the-job training as a means of training and socialization | Although informally important, clearly of secondary relevance. | The primary avenue by which neophytes enter the occupation. | Frequently reported as a critical form of training. The primary form of training in some technical occupations. |
| Evidence of formal occupational organization | Professional societies, licensing, accreditation boards, professional journals are nearly universal. | Unionization is common but not universal. | Mixed picure; some technical occupations have journals and professional societies, others have none. |
| Formal certification | Required to practice | Not required to practice | Common among technical occupations in medicine. |
| Other occupational means of controlling entry | High | Low, primarily through union control of apprenticeship programs. | Low, with the exception of technical occupations in medicine. |
| Tendency to unionize | Low | High | Less common than among crafts, more common than among the professions. |

by other technicians and related occupations, who might therefore be said to work in an occupationally structured milieu. However, it is possible to estimate the number of self-employed technicians using census data. Table 1.2 displays the percentage of individuals in each of the Bureau of Labor Statistics' nine general occupational categories who reported being self-employed at the time of the 1990 census. Only 2.5 percent of the technicians in the United States report being self-employed. In fact, the data on self-employment suggest that technicians are particularly likely to work in an organizational context: with the exception of clerical workers, no other occupational group is less likely to be self-employed.

At the same time, however, there is growing evidence that technicians may be less well integrated into a vertical division of labor than their employment context would otherwise suggest. A number of technicians' occupations have developed institutional forms similar to those of the established professions. For instance, radiological technologists and medical technicians have their own licensing procedures, occupational associations, journals and training programs. More important, even technicians whose occupations are less formally organized appear to orient themselves as much to their communities of practice as they do to the organizations by which they are employed. Synthesizing data on careers from a number of ethnographies of technicians' work, Zabusky and Barley (1996) report that few technicians desire hierarchical advancement and those who do typically cease to seek managerial duties once they have obtained supervisory responsibility for a small group of technicians. Instead, most technicians either remain in the same job for long periods of time or else move from organization to organization in search of greater technical challenge and respect for their abilities. Zabusky and Barley (1996) found that, like professionals and craftspeople, technicians almost always define career development in terms of knowledge, acumen, and skill. Technicians also argue that expertise can only be

TABLE 1.2
Percentage of self-employed individuals in various occupational categories

| *Occupational Cluster* | *Percentage Self-Employed* |
|---|---|
| Agriculture, forestry, fishing, and related occupations | 39.4 |
| Marketing and sales occupations | 13.0 |
| Executive, administrative, and managerial occupations | 12.8 |
| Precision production, craft and repair occupations | 11.9 |
| Professional occupations | 9.2 |
| Service occupations | 6.4 |
| Operators, fabricators, and laborers | 3.2 |
| Technicians and related support occupations | 2.5 |
| Administrative support occupations, including clerical | 1.5 |
| All occupations | 8.3 |

*Note:* Data are drawn from Table 6 in Silvestri and Lukasiewicz (1991).

accurately assessed by others in the field and that managers rarely understand technical issues even though they often have the authority to make or dispute technical decisions. Accordingly, technicians as diverse as microcomputer support technicians, medical technicians, science technicians and copier repair technicians routinely portray management and hierarchy as impediments to their work and argue that organizations rarely provide them with the respect they believe they deserve.

Thus, it seems that many technicians orient their identities and careers to the occupational rather than the organizational communities to which they belong; but because most technicians also articulate and subscribe to an ethos of professionalism that emphasizes responsible expertise, they rarely pit themselves against their employers (Nelsen and Barley 1994). Technicians may detach their identities from the organizations that employ them, and may even view managers and administrators as fools—but most remain deeply committed to the quality of the services they provide to whomever they consider to be their clients. For this reason, they often speak of wanting to contribute to the good of their employing organization, even as they distance themselves from the organization's system of reward and control.

### *Attempts at Resolving the Anomalies*

Although there have been few studies of how employers, educators, and unions perceive or treat technicians, existing evidence suggest that all are well aware of the problems that technicians pose for traditional conceptions of the workforce. Particularly salient has been the fact that technicians blur the boundary between mental and manual work. In the earliest investigation of employers' perceptions of technicians, James Brady et al. (1959) reported that some firms treated engineering technicians as if they were craftspeople, while others considered them professionals. Labor practices and employment relations varied accordingly. Brady et al. (1959) used such differences to justify their recommendation that scholars distinguish between "industrial" and "engineering" technicians. They claimed that although both types of technicians supported engineers, the former relied primarily on "manipulative skills" while the latter required "depth and understanding of engineering and scientific principles."

Whereas Brady et al. (1959) claimed that firms have tried to resolve the anomalies that technicians pose by merging technicians into either the blue-collar or the professional labor force, other studies indicate that some firms view technicians as "intermediate" workers. Summarizing interview data with executives of engineering firms, the authors of an early study of technicians in the Silicon Valley wrote: "Technicians have made their particular contribution as members of the professional team. They engage in work that requires some of the knowledge and skills of both the professional and skilled craftsman. . . . They have some 'know why' [theory] and 'know-how' [practice] and serve as a vital connecting link between the two groups. . . . Science and engineering technicians are semi-professionals . . . whose work requires competence in one of the fields of science or engineering and lies between that of the skilled craftsman and the professional" (California State Department of Education 1964, 10).

As the president of a small solid state physics company put it, "These science and engineering technicians, as you call them, are not new to industry, but they have created a new problem because they now occur in large enough numbers to be recognized as a group unto themselves" (California State Department of Education 1964, 13). Commentators who portray technicians as an intermediate group generally resist forcing technicians into either mental or manual occupational categories. However, they preserve the long-standing cultural image of a hierarchical relationship between mental and manual work by situating technicians between the two.

The U.S. labor movement has also generally dealt with technicians as if they were either craftspeople or professionals, in part because no American labor union exists specifically for technicians. Nevertheless, troubles arise regardless of the type of union with which technicians affiliate themselves. For instance, in 1957 the Engineers and Scientists of America (a now defunct association of engineers' and scientists' unions, founded in 1952) split over the role of technicians. One faction felt that allowing technicians into a bargaining unit compromised "professional objectives." Another argued that one could make no real distinction between professionals and technicians and that technicians were necessary if bargaining units were to have adequate strength (Brady et al. 1959). Technicians have usually fared no better when they have joined industrial or craft unions, where their professional orientation and their tendency to be more accepting of the organization's perspective places them in conflict with the larger membership. For example, in recent years some local chapters of the Communications Workers of America have actively excluded technicians from their bargaining units because they are perceived as having interests that differ from those of the rank and file.[12]

Educators who design curricula for the training of technicians have also wrestled with the technician's place in the division of labor in parallel terms. Melvin Barlow (1967) prefaced his discussion of how to design technical curricula by noting that technicians have more "manipulative skills" than engineers and scientists and more "technical skills" than craftspeople. Like employers who portray technicians as an "intermediate group," Barlow concludes that the mixture

12. This information was communicated during a discussion that occurred during a conference of CWA members held at the New York State School of Industrial and Labor Relations at Cornell University in October 1993.

of skills suggests "the approximate location of the technician in the hierarchy of occupations." More recently, Daniel Hull (1986) has proposed that educators adopt an almost Comptian view of the organization of technical work: "scientists" provide the basic knowledge that "engineers" use to design devices which "technicians" test and repair and "operators" run and monitor. In Hull's model, knowledge flows down a hierarchy of sophistication from scientists to operators as skills become increasingly manual. Since technicians' work lies somewhere between professional (mental) and blue-collar (manual) work, Hull advocates that technical schools design their curricula accordingly.

Managers and administrators of organizations that employ technicians also appear to be aware that technicians threaten the underlying premise of bureaucratic control by introducing elements of a horizontal division of labor into a vertically organized context. Zabusky and Barley (1996) found that administrators were often willing to admit that they could not easily evaluate the work of microcomputer support technicians because they lacked the necessary expertise. As a result, administrators tended to grant the technicians considerable operational autonomy. At the same time, however, they insisted that technicians could not be completely trusted because of their technical orientation. As one administrator put it, "They look at themselves as professionals and feel that they alone have the capacity to judge their work. But they can't really be regarded as professionals because they do not have a broader capacity for judgement." Similarly, Barley (1984) found that radiologists readily admitted that sonographers often knew as much as (or more than) they did about the interpretation of ultrasound images. As a result, radiologists generally treated sonographers with considerable respect. Nevertheless, the radiologists were quick to prohibit sonographers from providing referring physicians with interpretive information, lest the sonographers' expertise undercut their own position of authority in the traditional hierarchy of medical occupations.

Practitioners have sought to resolve the anomalies that technicians pose for traditional images of the workforce in much the same way that an earlier generation of sociologists sought to make sense of engineering. Technicians are either allocated to existing occupational categories or else portrayed as hybrids whose work is located in the interstices between those categories. Usually, attempts are also made to force technicians into a status hierarchy, which may be either organizational or occupational in nature. Given that employers have been unable to come to grips with the unique aspects of the technician's role, it is perhaps unsurprising that until recently most sociologists who have studied technicians have offered similarly conflicted interpretations of the technician's place in the division of labor. William Evan (1964), for instance, concluded that engineering technicians were "marginal" workers, while B. C. Roberts et al. (1972) argued that technicians were best viewed as "intermediate" workers.

## TOWARD A PRACTICE-BASED MODEL OF TECHNICIANS' WORK

The difficulty with such attempts to resolve technicians' anomalous status is that they all rely on concepts developed to characterize work in a previous era. A more promising strategy would be to develop a model of the technician's role based on the work that technicians perform. Historically, most lines of work have revolved around the manipulation of physical entities, the manipulation of symbols, and the management of people. As we have noted, the everyday meanings of "blue-collar" and "white-collar" reside largely in the gulf between these two foci. Managerial work is distinguished from other white-collar work because it also involves interpersonal communication or the manipulation of people (Zuboff 1988). Like engineering, technicians' work violates the cultural segregation of the material and the symbolic. However, unlike engineers, technicians have acquired little or no responsibility for people and are much less implicated in the control strategies of the firm. Moreover, as we shall see, technicians have a different relationship with the material world than do either engineers or craftspeople, and a different relationship to symbol systems than do professionals.

Studies have repeatedly shown that technicians work at the empirical interface between a world of physical objects and a world of symbolic representations. Using sophisticated technologies and techniques, technicians orchestrate the connection between the two. Technicians act as the link between a larger system of work and the materials on which the system depends. Depending on context, the materials of relevance may be hardware (Orr, 1990), software (Pentland 1991a), microorganisms (Scarselletta, Chapter 8), the human body (Nelsen and Barley 1993; Barley 1990), a manufacturing process (Hirschhorn 1984), or a variety of other physical systems. Similarly, depending on context, relevant representations may consist of data, test results, images, diagnoses, or even theories. As depicted in Figure 1.1, the technician's task at the empirical interface pivots on two complementary processes: transformation and caretaking.

On one hand, technicians employ technologies and techniques to transform aspects of the material world into symbolic representations which can be used for other purposes. For instance, technicians in medical settings produce images, counts, assays and other data useful for medical diagnoses (Barley 1990; Scarselletta, Chapter 8, Nelsen, Chapter 7). Technicians in nuclear power plants (Hirschhorn 1984) and other automated facilities (Zuboff 1988) create and monitor flows of information on production systems. Science technicians reduce physical phenomena to data or "inscriptions" from which scientists construct arguments, papers and grants (Latour and Woolgar 1979).

Yet technicians do more than generate symbols and information. Most are also responsible for taking care of the physical entities they oversee. Technicians are charged with ensuring that machines, organisms and other physical systems re-

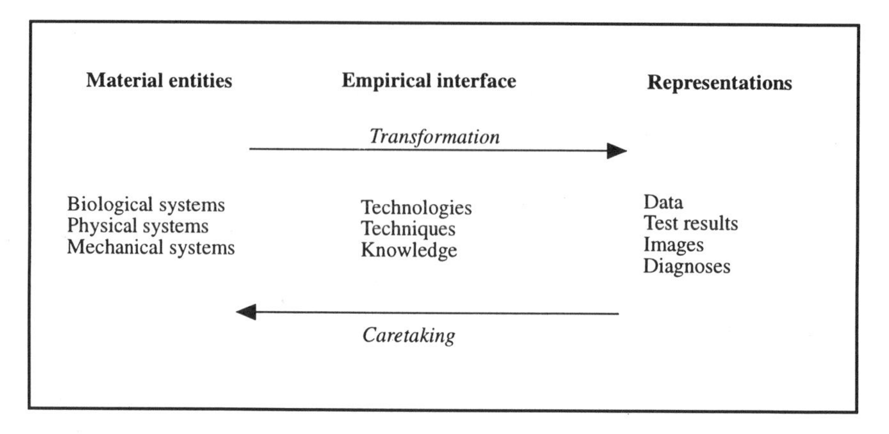


FIGURE 1.1
Technicians' work at the empirical interface

main intact and in good working order. Caretaking often requires that technicians employ the very representations they create. Thus, emergency medical technicians perform medical interventions based on diagnoses made at the site of an accident (Nelsen and Barley 1993). Technicians in biology labs employ the data they generate to husband organisms and monitor technologies (Barley and Bechky 1994). Microcomputer technicians use the results of tests and probes to alter the functioning of computer systems.

The dual processes of transformation and caretaking that define the core of technicians' work also generate its anomalous standing. To be effective, technicians must manage troubles at the empirical interface to ensure the smooth production of representations as well as the integrity of a physical system. To achieve this task, technicians must comprehend the principles of the technologies they employ and make use of abstract and systematic bodies of knowledge. In this respect, technicians resemble professionals. Science technicians, for instance, require knowledge of mathematics as well as knowledge of the science on which their practice is based (Barley and Bechky 1994). Emergency medical technicians, sonographers, and medical technologists require knowledge of biological systems, pharmacology, and disease processes to render diagnostic-ally useful information. Even microcomputer support technicians require an abstract understanding of how computers and software function. Without such formal knowledge, technicians cannot perform their caretaking function optimally.

But because technicians manipulate physical entities to achieve practical ends, they must also possess extensive contextual knowledge of materials, technologies, and techniques. Contextual knowledge is largely particularistic, acquired though practice and difficult to verbalize—much less codify (Pinch, Collins, and Carbone, Chapter 4). It resides in a practitioner's ability to find and interpret subtle visual, aural, and tactile cues where novices see no information at all. In this sense, technical work seems to resemble a craft. Craftspeople have long been valued for their ability to render skilled performances based on an intuitive feel for materials and techniques (Harper 1987; Becker 1978). Studies show that technicians pride themselves on, and are prized by others for, precisely the same sort of skill. For instance, Alberto Cambrosio and Peter Keating (1988) report that technicians in monoclonal antibody laboratories are often unable to fully articulate their techniques for producing viable hybridomas. Consequently, labs frequently cannot duplicate each other's work even when procedures are meticulously documented. The transfer of technical knowledge requires one laboratory to dispatch technicians to another, and even then, the recipient of the information may be unable to cultivate the cell line. The importance of contextual knowledge is a theme that runs though most studies of technicians' work (see Orr, 1996, Pentland 1991a, Scarselletta 1992, Collins 1974, Jordon and Lynch 1992).

The integration of formal and contextual knowledge in the course of manipulating objects and symbols simultaneously is precisely what makes technicians' work culturally anomalous. In this sense, technicians' work is neither a craft nor a profession and, hence, all such comparisons obscure the technician's role. Although technicians' work does involve operations on material entities, it is not a craft because technicians do not manipulate or transform materials to produce objects. Instead, they manipulate materials to produce symbolic representations. Technicians' work is not like that of traditional professions, despite the fact that technicians work with symbolic representations, because they do not use these representations to create other representations. Instead, they use representations to manipulate objects. Nor is technicians' work a hybrid in the sense that it is like a craft or blue-collar work on one dimension and like a profession or white-collar work on another. Instead, technicians' work punctures the existing cultural bulwark; it is at once a synthesis of mental and manual, clean and dirty, white-collar and blue-collar. Such a synthetic melding of cultural opposites has been previously approximated only by engineering, surgery, and other "manual" professions.

Similarly, technicians' work, like that of engineers, is not easily subsumed by either organizational or occupational images of structuring labor. Most technicians work in organizations that routinely attempt to fold technicians into a vertical division of labor via formal job classifications, graded pay scales, and related human resource practices. Yet technicians remain strongly oriented to their community of practice. More important, technical work routinely undermines bureaucratic systems of control and stipulated roles because technicians' expertise and knowledge is a critical resource which few outsiders possess. Thus, technicians' work brings the underpinnings of an occupational division of labor directly to the core of well-established organizational divisions of labor.

## CONCLUSION

What, then, are we to make of technicians, engineers, and technical work in general? Perhaps the most that can be said, at present, is that technical work creates conditions conducive to a more horizontal division of labor where expertise is balkanized into substantive domains that cannot be easily classified using only traditional images of mental and manual work. As such, technical work has posed a continuing challenge to the vertical divisions of labor that are the legacy of the Industrial Revolution. Such organizational structures ultimately triumphed over an earlier occupational division of labor by relegating manual work to the lower echelons of a hierarchy of authority and assigning symbolic and interpersonal work to the upper echelons. The feat was accomplished by separating execution from cognition and claiming that abstract knowledge provides a platform for directing the work of others. Much of engineering's professionalizing strategy has been an attempt to make itself into the organizational bearer of much of that abstract knowledge. New questions arise, however, concerning the future of technicians' work, which is developing in a very different sociocultural and economic environment from the one present during the rise of engineering.

In the long run, organizations may conceivably defuse the cultural challenge posed by technicians' work by absorbing technicians into the vertical division of labor. Attempts to treat technicians as either craftspeople or corporate professionals can be viewed as precisely such a move. It is also possible that technicians' work could serve for some as an apprenticeship to more fully professionalized positions, as has historically been the case in British engineering. However, it seems unlikely that organizations will be able to finesse the challenge posed by technicians in the same way that they once dealt with the anomalies created by engineering. Engineering could be more easily linked to management because the engineer's expertise in designing systems can be used to control the work of others. In contrast, most technicians' work is not about design, offers organizations no direct competitive advantage, and rarely carries responsibility for the work or well-being of people.

Firms transformed engineering into a corporate profession in part by offering engineers managerial careers, healthy salaries, and related benefits. In an era when many organizations are downsizing and flattening their hierarchies, firms are unlikely to possess the resources required to create internal labor markets and convincing loyalty packages for technicians. In fact, most technicians' jobs are currently divorced from the vacancy chains of their organizations. Instead, many technicians build their careers by seeking increasingly more challenging problems, a strategy which often requires that they move from organization to organization. Other technicians pursue careers by abandoning their current occupation to seek training in a related field (Zabusky and Barley 1996). For example, an appreciable number of emergency medical technicians eventually become police officers or firefighters, while a smaller number become nurses or doctors. Medical technologists either leave the field altogether or else migrate to other health care occupations.

Engineering's status as a corporate profession was also bolstered by its ties to the university. In contrast, most technicians' jobs require no more than an associate's degree from a community college or technical institute. Because associate's degrees have far less status than bachelor's degrees, it is unlikely that technicians will be able to use their formal education to justify the stature of corporate professionals unless society's image of the associate's degree changes.

Given these conditions, three scenarios for the future of technicians' work seem plausible. One possibility is that cultural understandings of mental and manual work may prove sufficiently malleable to mesh technicians' work into a vertical system for structuring labor and allocating status. Within such a scenario, technicians would most likely become the postindustrial analogue of the blue-collar worker. The fact that technicians are currently paid less well and accorded less status than their education and contributions warrant would seem to foreshadow such a resolution. The continuing decline of semi-skilled and unskilled work may also favor such a resolution, since the demise of blue-collar work would heighten the saliency of technical work's manual and contextual components. If technicians are transformed into the postindustrial equivalent of blue-collar workers, policy makers and employers should expect the current difficulty of convincing youth to invest in the education necessary for technical jobs to become even greater.

It is unlikely, however, that technicians' work could be structured in precisely the same way that semi-skilled and unskilled labor are structured in today's economy. Not only do the attitudes and orientations of technicians differ in important ways from those of blue-collar workers (Creighton and Hodson, Chapter 3), but the symbolic products of technicians' work typically feed the work of managers and professionals. Consequently, if technicians are to be folded into a vertical division of labor, the form of organization is less likely to resemble a traditional bureaucratic hierarchy than a hierarchy of occupational dominance. The resulting status structure would therefore be more akin to that of a hospital or scientific laboratory than that of a firm.

Alternately, the expansion of technical work may force organizations to return to a more horizontal division of labor. In such a world, technicians would be viewed as members of one of a series of occupations composed of experts whose knowledge and skills must be pooled with those of experts in other domains in order to accomplish a task. Technicians would then become members of multi-occupational teams. Preliminary evidence suggest that such team structures characterize the informal organization of everyday work in ultrasound (Barley 1990), histology labs (Scarselletta 1992), and scientific laboratories (Barley and Bechky 1994), even though the formal structures of these organizations still presume a more hierarchical relationship. One might view current interest in collaborative forms of organizing in industry as consistent with such a development. As knowledge becomes more extensive and differentiated, it becomes more difficult for management and members of other occupations to claim greater knowledge as a prerogative for control. Such structures may also provide solutions for engineering's potential disengagement from a close linkage with management career ladders in a world where the need for technical expertise may outrun the need for managerial expertise.

The third, and perhaps most likely, scenario is that different technicians' occupations will experience different futures. In the nineteenth century it would have been difficult to predict just which categories of technical and craft work would eventually become the functions of skilled craftspeople and which would be the ultimate responsibility of engineers and management. Indeed, where the boundary ultimately came to be drawn shows significant national variation (Whalley 1991a). We are in a similar situation today in predicting the future of the variety of tasks and activities we currently label technicians' work. Some technicians may find themselves treated like blue- or lower white collar labor, while others enjoy considerable occupational autonomy and recognition. Evidence for such a bifurcation of technicians' work can be found in radiology and pathology where sonographers, CT technicians, and histologists are generally granted greater respect for their expertise than are x-ray technologists and technicians in blood chemistry (Barley 1990; Scarselletta, Chapter 8). A key objective of future research on technical work must be determining the conditions that decide the technician's status and role in local divisions of labor.

Arthur Stinchcombe (1959) noted long ago that both organizational and occupational forms tend to exhibit characteristics peculiar to the historical conditions of their birth. On these grounds alone, there is reason to speculate that the social organization of technical work will not simply recapitulate earlier structures and cultures of work. What is at issue is how current means of organizing work will develop in a world increasingly composed of technical workers. The issue is ultimately an empirical matter whose resolution will become increasingly clear over time. Our goal has been simply to heighten awareness that the future of work may not be well served by images from the past.

# 2

# TECHNICAL DISSONANCE: CONFLICTING PORTRAITS OF TECHNICIANS

*Jeffrey Keefe and Denise Potosky*

Technicians are a rapidly growing occupational group of increasing importance to postindustrial society. The business press has championed technicians as the new worker elite, and politicians predict that technicians will form the core of a new postindustrial middle class. Technicians' occupations, however, are generally not well understood. The research literature that does exist offers highly conflicting portraits of technicians as postindustrial factory operatives, as modern successors to skilled craftworkers, and as emergent professionals. The research reported in this paper attempts to bring these multiple images into focus.

The previous chapter, by Peter Whalley and Stephen Barley, concluded that there are several core characteristics shared by most technicians' occupations. Technicians' responsibilities are esoteric and skilled in nature, and often involve working with or on complex technologies and technical processes. Their work requires the sophisticated manipulation of symbols and data and the diagnosis of problems.

Most technicians' occupations are also characterized by the acquisition of a craft version of professional knowledge (Abbott 1988, 65). As technicians' occupations are institutionalized, some elements of formal scientific, technical, or professional training are commonly required (Carnevale, Gainer, and Schulz 1990). It is this linkage of skilled practice to a formalized knowledge system that distinguishes the technician from the skilled craftworker. Although often linked to a scientific or professional knowledge system, the technicians' skilled practices often remain tacit and contextually specific, not formally "scientific" (Barley and Bechky 1994).

Technicians' occupations are also uniquely passive in their social construction. Whereas the professions and crafts are often constructed by the active social organizing of their prospective incumbents, technicians' occupations are largely constructed and defined by more powerful others. A technicians' occupation is often the by-product of the integration of a technical profession into large corporate or public enterprises. The new occupation is assembled from tasks discarded by professionals through the "hiving off" process described by Whalley and Barley.

As a consequence of the hiving off process, most technicians work in complex organizations where they neither set the entry and performance standards of their occupation, nor control the educational process through which new recruits are trained. Most cannot formally self-regulate or self-govern their work practices. Most important, technicians often operate within an established profession's field of knowledge and competence (Abbott 1988). The result of performing their skilled work within an established profession's jurisdiction is a blurred identity. The allied profession is socially acknowledged to control the technicians' entire knowledge system, whether theoretical or contextual. As a result, technicians are highly skilled workers who lack any formal control over their own occupations; they work within the orbit of a dominant allied profession (Trice 1993).

In the formation of technician's occupations, the process of hiving off has followed two models of development; one is organizational, the other occupational. The distinction between these models depends on whether technicians substituted for or complemented scientists or other professionals in the specific field. For example, science technicians, as a distinct occupational group, appeared at the historical moment when science was bureaucratized and relocated into large public and corporate laboratories. As substitutes in the public health laboratories after 1895, technicians performed standardized diagnostic testing so skilled bacteriologists could devote their time to research on infectious diseases (Gossel 1992). In the testing laboratories, technicians were substituted for scientists and were thus integrated into an organizational division of labor.

With the formation of industrial research laboratories, such as GE (1900), Du Pont (1902), and AT&T (1911), first-rate university-based scientists were often promised the assistance of technicians as a recruiting enticement, along with higher pay, better equipment, and more research time (Reich 1985; Hounshell and Smith 1988). By employing technicians to conduct the more routine and physical aspects of their research, scientists could devote more time to theoretical endeavors (Roberts 1983). In this setting, technicians served to support and complement the work of scientists; consequently they were integrated into an occupational division of labor.

The technician in the industrial research laboratory remained closely tied to the scientist in an occupational division of labor. The scientist possessed authority over the technician and defined the nature, scope, and meaning of the technician's labor (Shapin 1989, 562). Technicians became largely invisible because they worked within an occupational division of labor where a profession was acknowledged to control their entire knowledge system. Both the theoretical and practical skills of technicians were within the profession's jurisdiction, which denied them an independent identity. Technicians became visible within the organization only when they made mistakes, departed from their assigned routines, or demonstrated incompetence (Shapin 1989, 558).

In the first section of this paper we use *Current Population Survey* data to analyze, compare, and contrast technicians, technical professionals, and other technical workers to assess how closely technicians' occupations orbit the technical professions. In the second section we describe two groups of technicians within a single company. One group works in a manufacturing organization and the other is employed in an occupational division of labor. We also compare both groups of technicians to skilled tradesworkers in this same company. This comparison allows us to explore how each group of these skilled technical workers attempts to control its own training, education, skills, work practices, and its relationship with management.

Management efforts to increase the technization of the skilled trades by upgrading their educational requirements in this company have encountered substantial resistance. As a group, technicians have been found to be dissatisfied with their status and treatment within the firm (Roberts et al. 1972). They have also been found to be dissatisfied in terms of their autonomy, their career opportunities, and their opportunities for self-development, and as individuals. Professional career expectations formed in educational programs have rarely conformed to their job experiences.

Prior research identifies status incongruity and role ambiguity as major sources of technicians' dissatisfaction and dissonance leading to widespread feelings of deprivation (Roberts et al. 1972, 36). The occupational roles of these technicians were often defined by the professionals with whom they worked. Technicians' invidious comparisons with professionals (Evan 1964) and high levels of status inconsistency were correlated with expressed dissatisfaction, increased levels of interest in unionization, heightened interest in occupational certification, and reduced occupational and organizational commitment (Koch 1977). Status conflict, role ambiguity, invidious comparisons and their associated dissatisfaction and dissonance are thought to arise because technicians often operate in conjunction with an established profession. Prior research has compared technicians' job satisfaction and status with professionals; in the last section of this chapter we use survey results to examine their job satisfaction in relation to the satisfaction of production workers and skilled tradesworkers to question how dissatisfied technicians actually are with their status and power.

In the research reported in this paper, we have used three distinct research methodologies to help focus these multiple images of technicians: archival data analysis, case study, and attitude survey. Our research effort is exploratory and intended to be hypothesis-generating.

## TECHNICIANS WITHIN THE TECHNICAL OCCUPATIONAL STRUCTURE: COMPARATIVE PATTERNS

By examining commonalities, differences, and changes among and within professional, craft, and operative technical occupations (using *Current Population Survey* data from 1983 and 1993), we uncover some characteristic patterns that help clarify the relationships, connections, and associations of technicians with other technical occupational groups. Overall, the data on employment growth, wage growth, gender composition, and unionization strongly indicate a high level of interdependence and interconnection between technicians and technical professionals.

A total of four million working technicians accounted for 3 percent of US employment in 1993 (Table 2.1). Between 1983 and 1993, approximately one million new technician jobs were created, increasing technician employment by 31 percent, a rate 70 percent faster than overall U.S. employment growth (18 percent). Only employment of technical professionals grew at a faster pace (36 percent), which was double the rate of U.S. labor force growth during the decade 1983 to 1993. In aggregate and within more detailed occupational categories, professional employment grew faster than technician employment; however, the patterns suggest that the growth of professional and technicians' occupations are linked. Engineers and engineering technicians posted the slowest rates of employment growth (11 and 6 percent respectively), while employment of health care professionals and health technicians each grew at 37 percent. Technicians were also less numerous than technical professionals, representing only three-fifths the number of technical professionals in 1983 and slightly less than three-fifths in 1993.

Technicians closely reflected the gender and racial composition of the overall labor force in 1993. Technicians were 51 percent female and 10 percent black in 1993, as compared to the U.S. labor force, which was 46 percent female and 10 percent black. While technical professionals also resembled the labor force in gender composition (45 percent female), the professions employed blacks (5 percent) at half the rate at which they were found in the labor force. On the other hand, the skilled blue collar occupations of construction trades (2 percent female) and mechanics and repairers (3 percent female) remained almost exclusively male domains; they have made some progress in increasing black participation

TABLE 2.1
Changing composition of employment among technical occupations, 1983–1993

| *Technical Operations (Numbers in Thousands)* | *1983* | *1993* | *Percentage Change* |
|---|---|---|---|
| Technical Professionals | 5,130 | 6,952 | 36 |
| Engineers, surveyors, architects | 1,675 | 1,859 | 11 |
| Math and computer scientists | 463 | 1,051 | 127 |
| Natural scientists | 357 | 531 | 49 |
| Health diagnosing | 735 | 909 | 24 |
| Health treating | 1,900 | 2,602 | 37 |
| Technicians | 3,053 | 4,014 | 31 |
| Engineering technicians | 822 | 870 | 6 |
| Computer programmers | 443 | 578 | 30 |
| Science technicians | 202 | 261 | 29 |
| Health technicians | 1,111 | 1,522 | 37 |
| Mechanics & repairers | 4,158 | 4,416 | 6 |
| Construction trades | 4,289 | 5,004 | 17 |
| Precision production | 3,685 | 3,758 | 2 |
| Total skilled technical occupations | 20,315 | 24,144 | 19 |
| Machine operators & assemblers | 7,744 | 7,415 | −4 |
| Transportation operators | 4,201 | 5,004 | 19 |
| Total skilled technical operators | 11,945 | 12,419 | 4 |
| Total U.S. labor force | 100,834 | 119,306 | 18 |
| Percent of labor force in: | | | |
| Skilled technical occupations | 20% | 20% | |
| Semi-skilled technical occupations | 12% | 10% | |

*Note:* Sources for Occupational Employment:
1983: *Handbook of Labor Statistics* (Bulletin 2340 USDOL 1989) Table 18.
1983: *Employment and Earnings* (Jan 1994, USDOL) Table 22.

(7 percent in each category), but without much improvement in the share of black employment during the last decade. Thirty years after Title VII, gender exclusion remained a core characteristic of traditional skilled trades occupations and their progeny. This record, however, does not compare unfavorably with the largest technical profession, engineering, which was only 9 percent female and 4 percent minority in 1993.

Technicians joined unions in increasing numbers in the 1970s, as reflected in the growth of union coverage for engineering and science technicians from 16 percent in 1974 to 19 percent in 1980 (Kokklenberg and Sockell 1985). Most of this growth occurred in the public sector. The pervasive decline of unionization in the 1980s was shared by all technical occupations (Curme, Hirsch, and MacPherson 1990). Unionization of technicians fell from 19 percent in 1983 to 12 percent in 1991, while unionization among precision production, craft, and repair workers went from 36 percent to 26 percent in the same period. Professionals were the group least affected by declining unionization, going from 30 percent to 26 percent union coverage between 1983 and 1991. However, union coverage of technical professionals was only 13 percent in 1991. The experiences, trends, and levels of unionization of technical professionals and technicians are highly similar. Both groups have unionized at about one half the rate of skilled and semi-skilled blue-collar workers and non-technical professionals.

Technical professionals experienced rapid real wage growth between 1983 and 1993. During this period, real wages increased 20 percent, with large increases for health diagnosis (51 percent) and health treating (30 percent) professions (Table 2.2). Technicians, on the other hand, experienced a modest 3 percent real wage gain for this decade, with real wage decreases for engineering (−4 percent) and science (−9 percent) technicians and increases for computer programmers (13 percent) and health technicians (11 percent). That is, technicians' wage growth failed to keep pace with technical professionals in any specialized occu-

TABLE 2.2
Median weekly earnings and real wage changes for technical occupations, 1983–1993

| *Technical Operations* | *1983 Wages* | *1993 Wages* | *Real Wage Change* |
|---|---|---|---|
| Technical Professionals | $497 | $818 | 20% |
| Engineers, surveyors, architects | $594 | $902 | 7% |
| Math and computer scientists | $542 | $895 | 20% |
| Natural scientists | $512 | $722 | −4% |
| Health diagnosing | $506 | $994 | 51% |
| Health treating | $393 | $687 | 30% |
| Technicians | $357 | $528 | 3% |
| Engineering technicians | $390 | $550 | −4% |
| Computer programmers | $473 | $747 | 13% |
| Science technicians | $369 | $501 | −9% |
| Health technicians | $294 | $458 | 11% |
| Mechanics & repairers | $376 | $504 | −11% |
| Construction trades | $374 | $495 | −13% |
| Precision production | $378 | $490 | −15% |
| Total skilled technical occupations | $404 | $595 | 2% |
| Machine operators & assemblers | $262 | $333 | −18% |
| Transportation operators | $328 | $447 | −9% |
| Total skilled technical operators | $286 | $379 | −13% |
| Total U.S. labor force | $313 | $463 | 3% |

*Note:* Sources for Occupational Employment:
1983: *Handbook of Labor Statistics* (Bulletin 2340 USDOL 1989) Table 43.
1983: *Employment and Earnings* (Jan. 1994, USDOL) Table 56.
CPI Deflator: *Economic Report of the President*, 1994, Table B-59.

pational category. Consequently, the wage gap between technicians and technical professionals grew during the decade. Technicians earned 72 percent of the professional's weekly earnings in 1983; however, by 1993 they received only 65 percent of professional earnings. By keeping pace with inflation, however, technicians' earnings significantly out-performed those of both skilled and semi-skilled blue collar technical workers, who experienced real wage declines ranging from 11 to 18 percent. In 1983, technicians earned 5 percent less than did skilled blue-collar technical workers, but by 1993 they earned 5 percent more than these workers. The 5 percent higher median pay received by technicians lends further support to the claim that technicians are an increasingly important skilled occupational group.

In terms of education, technicians, more than any other occupational group, were required to possess a two-year associate's degree as a minimum qualification, according to data collected in a 1984 special supplement to the Current Population Survey. According to this supplement, 24 percent of technicians needed at least a four-year bachelor's degree and 20 percent needed, at minimum, a two-year associate's degree. Since two-year vocational degrees were still relatively new in 1984, we believe it likely that this requirement has expanded over the last decade. In contrast to the 24 percent figure reported for technicians, almost two-thirds of technical professionals were required to have earned a bachelor's or a more advanced degree in order to qualify for their present job, and 12 percent of technical professionals needed an associate's degree. While relatively few blue-collar skilled or semi-skilled workers were required to have a bachelor's degree, approximately 5 percent of the skilled blue-collar technical workforce needed an associate's degree to qualify for their positions, which was one quarter the rate required of technicians.

The data on both real wage growth and qualifying education suggest that technicians may be viewed as an intermediate occupational category situated between technical professionals and skilled blue-collar workers. The growing identification of technicians with two-year associate's degrees suggests a possible institutionalization of this unique intermediate position, identifying them as members of neither a craft nor a profession.

Although technicians lagged behind technical professionals in both employment and wage growth over the last decade, the patterns of growth show a high degree of association at both the aggregate and more detailed occupational levels. The more rapid employment and earnings growth of professionals does not support the notion that technicians serve as a lower-skilled and lower-waged substitute for professional labor. In aggregate and within each specialized occupational group, the expansion of professional employment and increase in earnings exceeded the growth rates for technicians from 1983 to 1993. A more likely explanation of this pattern is that technicians served to complement and support technical professionals in the technical division of labor. This explanation suggests that the demand for technicians is derived at least in part from the demand for technical professionals.

Taken together, this data can be used to construct a composite sketch of technicians as an occupational group that combines the attributes of both the skilled craftworker and the professional. Their intermediate position may reflect an occupational group in the process of institutionalizing its unique identity. This optimistic image is widely reported in the popular press. For example, one recent report in *Fortune* magazine described technicians as the new "worker elite" who have become the core employees of the information age. As a result of this new status, technicians' occupations have allegedly become the new anchor for people's careers (Richman 1994, 56).

Yet there remains some evidence not consonant with this portrayal. Technician's wages closely matched those of skilled blue-collar workers, and the wage gap between technicians and technical professionals has widened. Most notably missing from these popular images of the new "worker elite" are two attributes central to the definition of a craft or a profession: occupational members set entry and performance standards and individual members engage in self-regulation with considerable autonomy in their work practices (Hall 1968; Freidson 1986). Neither is characteristic of technicians' occupations. Instead, technicians may be subordinated within complex divisions of labor, either organizational or occupational (Freidson 1977), without any real institutionalized power over their occupation or their work (Simpson 1985). This situation may give rise to their widely reported dissatisfaction. The macro data, however, cannot adequately explain contextual questions of occupational control and autonomy. To examine these issues we conducted to a case study at a large pharmaceutical company in New Jersey.

## A CASE OF ANALYSIS COMPARING TECHNICIANS AND SKILLED TRADESWORKERS

A large pharmaceutical company is an ideal setting for the study of technicians. Chemicals and related industries, such as pharmaceuticals, employ 19 percent of the nation's quarter million science technicians.

At the request of the Joint Labor-Management Apprenticeship Council, we developed a survey covering a variety of work-related attitudes, including job satisfaction, work characteristics, and preferences for different types of work within the organization. We began our project in September 1992 and completed it in August 1993. During this period we conducted over forty interviews with individuals at all levels within the organization, held several focus groups, attended a variety of meetings, and surveyed 337 workers, including forty-three laboratory technicians, ninety-six skilled tradesworkers, and 198 semi-skilled production workers. The information obtained in the interviews, focus groups, and meetings form the basis of our case analysis. The results of the survey are presented in the last section of this chapter.

The workforce we studied was approximately 60 percent female and 30 percent minority. The technicians, skilled tradesworkers, and other workers were all represented by an industrial union with a single local at this location. In 1993, this facility employed 190 skilled tradesworkers in its maintenance department. These workers were members of a wide range of occupations, including the construction trades, such as electricians, carpenters, and pipe fitters; mechanics and repairers, such as millwrights, instrument repairers, refrigeration mechanics, and various specialized industrial machinery mechanics; and precision production workers, such as machinists. Another 650 workers were employed in manufacturing. Approximately 125 science technicians were employed in the Quality Control (QC) and Research and Development (R&D) laboratories at this site. Ninety of these were chemical technicians; thirty-five were biological technicians.

### *Diversity in Lab Work of the Quality Control and R&D Laboratories*

The QC labs were an integral component of the company's manufacturing organization. They were responsible for testing raw materials, drug containers, and manufactured pharmaceuticals. Under FDA regulations, the company was required to track the manufacture and testing of each batch of pharmaceuticals it made. Working from a batch number, the company's records indicated when a drug was made, when the raw materials were received, who tested the raw materials, and when they were tested. The records provided similar information for containers. Records also documented who tested the manufactured drug, what tests were performed and what results were obtained, and where and when the drug was tested. A sample from each batch of pharmaceuticals was kept in a controlled environment for a specified period of time after its manufacture. This highly integrated information system required each QC technician to follow strict standardized testing and documentation procedures. Technicians were expected to document their tests precisely. Consequently, quality control technicians performed repetitive, carefully prescribed procedures, known as "cookbook science."

In the course of the last ten years, electronic automation had dramatically changed QC testing. Technicians no longer counted, weighed, or measured; machines performed these tasks, along with many others. Monitoring and maintaining the expensive electronic and testing equipment had become a major part of a technician's job. If there were testing problems, technicians needed to be able to determine whether the trouble was in the equipment or in the quality of the material being tested. Most of their work, however, was routine and boring. For this reason, several people interviewed commented that the QC lab techs had the worst jobs on the site.

The technicians who worked in R&D were at the bottom of a highly credentialed hierarchy. Scientists held doctorates; associate scientists had either M.S. or B.S. degrees, and new lab techs at a minimum held associate's degrees. R&D, unlike QC, occasionally hired individuals with bachelor's degrees into the bargaining unit to be lab techs. Although there was a highly formal status and authority hierarchy in R&D, lines of work were much less clearly defined than in QC. Work in R&D was considerably more flexible. Like lab work in QC, R&D lab work required a considerable amount of documentation, since successful projects might eventually lead to FDA drug trials. However, R&D permitted a much greater diversity of work experience for technicians than did QC. Technicians could assume more responsibility, if they wanted it. The company, however, could hold them accountable only for the practices covered in their job descriptions. While many tasks were mundane, R&D technicians might be asked to work on more interesting tasks such as instrumentation or experimental profiles.

In R&D, technicians could work one-on-one with a scientist or in project teams. Most often, technicians were assigned to scientists, and, as one company representative explained, the scientists "wanted flexible people who [didn't] complain." Thus, successful work relations often depended on the ability of the technicians to get along with the scientists.

Neither the company nor the union made any formal distinction between QC and R&D technicians in terms of job title (they were all "technicians") or pay level. Yet substantial differences existed between QC and R&D technicians in terms of supervision (by lab supervisors vs. by scientists), the nature of their work (equipment monitoring and troubleshooting vs. complementing the work of scientists), and the quality of the work environment (routine interface with equipment and supervisors vs. interaction within project teams or one-on-one with scientists). Many R&D technicians' jobs required skilled practice in a variable and changing environment often highly dependent on their social interactions with scientists. The QC technicians' jobs did not. Their work systems in many ways resembled those of production workers. The quality control technicians were fully subordinated and integrated into a manufacturing hierarchy which required repetitive, precise, and strictly regulated testing synchronized to the rhythm of automated production. In order to understand these differences better, it is useful to examine the training and work history of technicians relative to that of the skilled tradesworkers in this organization.

### *A Case History of Apprenticeship Training for Technicians and Skilled Tradesworkers*

Over the last two decades, the training programs for both technicians and skilled tradesworkers underwent major restructuring, which remained a source of ongoing tension between the company and union. Controversies over the appropriate domains of occupational knowledge and skill have centered around the issues of selection criteria and educational requirements for each of these job titles.

Until the mid-1970s both science technicians and skilled tradesworkers were trained in federally registered apprenticeship training programs. These four-year programs combined evening study in vocational and technical high school courses with a series of on-the-job demonstrations. In the course of the four-year apprenticeship, a lab tech or skilled trades apprentice was required to demonstrate his or her competence in 100 to 150 tasks before being certified as a "journeyman" or "technician."

The apprenticeship program provided an opportunity for upward mobility for semi- skilled production workers. Whenever possible, technician and skilled trades apprentices were drawn from the production workforce. The most senior qualified bidders were selected.

Prior to 1972, eligibility for apprentice training was determined by seniority, age, physical condition, and a written qualifying test. The maximum age for apprentice applicants was forty years. Candidates had to undergo a medical examination by the company doctor; and most important, they had to pass a vocational pre-test administered by the county apprentice coordinator for vocational and technical high schools. The test examined the candidates' mechanical aptitude and basic math and reading ability. The test, however, was never validated.

In the early 1970s, the federal Bureau of Apprenticeship and Training (BAT) informed the company that the test would be indefensible if it were challenged in an Equal Employment Opportunity complaint. The company immediately eliminated the test and age requirements as selection criteria for apprenticeship training. This left seniority and the company physical as the only legally sanctioned criteria for selecting apprentice candidates. Yet the industrial psychology literature has demonstrated that seniority does not predict performance either in training or on the job. From one manager's perspective, the new selection criterion was reduced to whether the senior bidder had a "warm body." They reported that apprentice quality started on a downward spiral. Throughout the program, not only at school but in the workplace as well, apprentice training standards for both technicians and skilled tradesworkers were relaxed. In the vocational and technical school, no one ever failed. On the job, everyone was passed through their required tasks, partly because the company was rapidly expanding and needed technicians and skilled tradesworkers. The federal government, however, soon forced management to recognize that the labs needed to improve the performance of their lab technicians dramatically.

### *Problems with the Quality of the Quality Control Lab Technicians*

In the mid-1970s, the company encountered some difficulties with the Food and Drug Administration (FDA), which regulates the quality standards and procedures for the manufacture of pharmaceuticals and approves new drugs. The company's problems with the FDA led to a shake-up in the management of the organization. A new vice president of quality control, a scientist with a Ph.D., was hired. An internal investigation initiated by the new vice president revealed an extremely high error rate in basic laboratory activities, such as measuring, weighing, and counting. The investigation concluded that the laboratory technicians were the primary source of the company's quality control problems. Investigators found that many quality control technicians were poorly selected and trained, irresponsible, and unmotivated. Technicians often did not understand the quality procedures they were implementing. They could not explain why they were doing specific tasks or tests. As one manager explained, management reached the consensus that the company needed better people in the labs— "it could no longer tolerate mediocrity." A management committee was established to explore alternatives for selecting and training laboratory technicians.

### *The Professionalization of Lab Technicians*

After weighing alternatives, the management committee concluded that upgrading could be best accomplished by recruiting technicians with associate's degrees. The local community colleges had established two-year programs for chemistry and biology technicians that were already producing highly qualified graduates for hospitals and other companies in the area. To implement this change, however, management would have to persuade the union to relinquish the laboratory technician apprenticeship program. Management argued that achieving a first-rate quality control operation was essential to maintaining its production at the facility. This argument was advanced at a time when pharmaceutical companies were being lured from New Jersey to Puerto Rico by extremely favorable changes in federal tax laws, a situation which made the union highly responsive to the company's concerns.

After a series of quid pro quos were bargained, the union agreed to abolish the technician apprenticeship program. The change was ratified by the local membership in a contract vote. In return, the company would pay all educational costs for any union-represented worker who wanted to become a lab tech by attending an approved community college evening program. Like an apprenticeship, the program could be completed within four years. All lab tech vacancies would be posted internally. The company obligated itself to select the most senior incumbent employee, holding an appropriate associate's degree, before hiring externally. Since the abolition of the apprenticeship program, however, less than half a dozen incumbent workers have earned degrees to qualify for the more than forty lab tech vacancies. All other candidates have been hired from the colleges or from other employers. The new degree requirement effectively eliminated one upward track in the company's internal labor market for production workers.

The company relied on the community college programs to screen and provide basic training for technicians. Enrollment screening at the colleges was accomplished by candidates taking a standardized basic skills test, which predicted success in the community college program. Further screening occurred through course grades. The degree programs required candidates to take classes in science theory and math, laboratory courses, and several liberal arts courses, including English. Community college programs also fostered an appetite for professional status among the technicians. The graduate technicians arrived at the company with high personal aspirations and a basic understanding of laboratory science. Supervisors and scientists were then formally responsible for ensuring that technicians were trained in their specific jobs. The community college educated lab technicians told us that their college labs and the math and English classes were useful to them on the job. The theory courses in chemistry and biology, however, were not considered relevant, since technicians were not responsible for developing research methods for the company. When we asked how they learned to do their jobs, the technicians initially recited the company's formal training procedures, which relied on the community colleges for basic education and held supervisors and scientists responsible for training lab techs on the job. With some probing, however, they indicated that other lab techs, and not the supervisors, scientists, or college instructors, had actually trained them.

As management had hoped, the community college-trained lab technicians did improve quality control. These lab technicians understood the importance of precisely followed procedures and detailed documentation. Encouraged by their success, management decided to recruit people with more advanced degrees to be technicians. If an associate's degree was good, then logically, a bachelor's degree would be better; a master's better still; and a doctorate would be best. Because of the company's reputation as a good employer and the union pay scale, management was able to implement its upgrading plan. The company even recruited Ph.D.'s, mainly foreign-born scientists, into the bargaining unit. This plan proved disastrous. In Quality Control these more advanced degree-holders found their work boring, routine, and even demeaning. Once they realized that it would take at least three to five years to get a promotion, many of them expressed their intense displeasure with everyone, and many quit.

Learning from this experience, by 1993, Quality Control wanted only two-year degree technicians. Quality Control did have a limited number of professional positions within the bargaining unit, such as analytical chemist, bacteriologist, and microbiologist. Each of these job titles had two levels, one for B.S. graduates, the other for M.S. graduates. Supervisors in Quality Control held at least a bachelor's degree and were usually promoted from the professional ranks of analytical chemists, bacteriologists, and microbiologists. Since most lab techs possessed at most an associate's degree, this practice of promoting only those with more advanced degrees effectively eliminated any upward mobility for technicians and effectively denied them any formal decision making power or control over their occupation.

Management's investigation of the QC labs' quality problem revealed that lab techs were not performing even basic laboratory activities, such as measuring, weighing, and counting, accurately. They attributed this problem to poor selection, training, and motivation of apprentice-trained lab techs. Once diagnosed, the problem was readily solved by seeking a new source of supply. In a labor market with an excess supply of educated labor (Freeman 1976), management's demand for community college-trained technicians was relatively easy and inexpensive to fill. Yet, did it solve the difficulty or was it, possibly, a solution looking for a problem?

### *The Framing of Quality Control's Quality Problem Does Not Yield a Quality Solution*

If the apprentice-trained lab techs failed in their basic tasks, management's logic suggested, they were incapable of performing more complex activities. However, if lab techs were not accurately measuring, weighing, and counting in the QC lab, was this behavior tolerated in production workers, the source supply of the apprentice-trained lab techs? We found no evidence to suggest that it was.

One decade later this same quality problem probably would have been solved using Total Quality Management (TQM) techniques. Analogous to Frederick Taylor's 1911 analysis of productivity, TQM argues that quality is fundamentally a management problem. It identifies only two possible sources of employee mistakes: lack of knowledge and lack of attention. Lack of knowledge can be easily solved by training while lack of attention "must be corrected by the person himself or herself, through an acute reappraisal of his or her moral values" (Crosby 1985, 83). If reappraisal fails, then employee discipline and discharge remain as the only viable solution. From the TQM perspective, the QC lab techs did not suffer from lack of knowledge, but from a lack of attention. Education and training did not comprise the prescribed solution for a lack of attention. At its root cause was a moral question of whether or not lab techs were going to ensure that the manufactured pharmaceuticals conformed to established quality control standards. According to TQM, it was management's job to guarantee that their answer was conformance.

Instead, management in the 1970s chose credentials over conformance. By recruiting community college-trained lab techs into this production environment to solve its quality problem, however, management engendered a new set of difficulties. In response to its quality control issues, management attempted to replace apprentice-trained technicians with college-trained technicians. This effort ordinarily might be expected to professionalize the technicians' ranks. In the QC labs, however, the new techs were integrated into an industrial bureaucracy which with the advance of technology brought about a greater routinization of QC work. Eventually, automation eliminated the potential for human errors in measuring, weighing, and counting. In R&D, the new technicians were formally slotted into the bottom of R&D's occupational hierarchy.

### *Technization of the Skilled Trades*

In the early 1980s, management once again decided to address a problem by relying on community college programs, this time for skilled tradesworkers. However, the outcome of this effort was vastly different from what happened to the technicians.

The rapid diffusion of new electronic controls prompted the company to demand a change in the apprenticeship program. The journeymen who would be responsible for maintaining electronic controls and the instrument repairers were only skilled in working on pneumatic control devices. They had received no training in electronics. From management's perspective, the trades apprenticeship program was also on the same downward spiral that had caused alarm about the quality of lab tech training. In the past, management had said that they could tolerate some poorly trained journeymen, because they could be assigned routine or "scut" work. The new technology, however, was rapidly eliminating such work.

Managers also noted a change in the motivations of the apprentices. Traditionally, apprentice candidates had often wanted to learn a specific trade; now, the candidates were primarily interested in making more money and did not care what trade they worked in. This indifference to one's trade, according to management, meant that these apprentices were less motivated to learn at school and on the job. The program's major problem was that it was unable to screen out these less motivated apprentices. For the traditional program to work, it needed highly motivated, self-disciplined apprentices. Managers felt that they had lost control over the selection and screening of apprentices, because of the Equal Employment Opportunity (EEO) required elimination of the test and because of a steady decline in the educational quality of vocational and technical schools. As technology advanced, the vocational and technical high schools were unable to keep up, further exacerbating management's disaffection. As one manager explained, when the company wanted to change the instrument repair apprentice program from pneumatics to electronics, the "voc-tech had nothing to offer us."

A management committee was established, under the leadership of the same manager, an engineer by training, who had spearheaded the changes in the lab tech program. Management decided, once again, that the vocational and technical high school program represented a major obstacle to improving the quality of the skilled trades apprenticeship program. The company informed the union that it would not post any new apprentice openings and threatened to begin subcontracting the work unless progress was made on the school quality issue. To solve the immediate problem of getting electronic instruments training, management proposed that the course work be moved to a well-respected technical institute in the area. In 1984, the union relented, after being promised that several new apprentice job openings would be posted. At a membership meeting, the local union ratified the changes in the apprenticeship program. Shortly thereafter, management proposed modifications in school requirements for each of the other trades' programs. This time the course work moved from the voc-tech to the local community college. Again, new apprentice openings were exchanged for each trade's concession to relocate their formal training to the college.

In order to attend the community college, the apprentice candidates were required to pass the New Jersey Basic Skills Test (BST). This test was designed to predict success in college. To bid on an apprentice job opening, a candidate had to have taken and passed the BST, which provided a legally sanctioned screen on job bidders. Once admitted, each trade's college program required an apprentice to take ten or eleven courses over a four-year period. One exception existed for the instrument repair apprentices, who continued to attend the technical institute.

From its inception in 1984, to 1993, the new apprenticeship program graduated eighteen journeymen. In 1993, six apprentices were enrolled. Six other apprentice candidates had left the program before graduation: three were "kicked out" when they failed college courses, two bid out to non-trades jobs because they were not passing their task requirements, and one was bumped out in a layoff. The new program had a 25 percent attrition rate even after using the Basic Skills Test as an admission screen. Management believed that the new program had solved both the apprentice quality issue and advanced technology education problems. This point of view was not shared by the skilled trades, who unanimously opposed the new program. In order to understand the skilled trades' opposition to the community college based formal education, it is necessary to consider the program it replaced.

### *Why the Skilled Trades Opposed the Community College Program*

In the voc-tech program, experienced tradespeople taught virtually all the courses. These teachers could easily relate the course material to work practices. Classes met twice a week, one night in a classroom, the other in a workshop. Apprentices were encouraged to speak to their instructors about any technical problems they were having on the job. The school programs also helped socialize the apprentices into their trade. The programs enrolled students from a wide variety of workplaces. Apprentices quickly learned that their new trade provided them with a multitude of employment opportunities and could guarantee their security and mobility.

On the job, apprentices were assigned to journeymen; they often started as helpers. As an apprentice gained experience, the journeyman trained him or her in a variety of tasks. When the apprentice was ready, a supervisor, who was a former skilled tradesworker, evaluated the apprentice's performance and certified the task's satisfactory completion.

Trades people believed that the only legitimate test for the apprentice was whether he or she could do the job. Consequently, they believed that formal course work should be aimed only at improving job performance. The journeymen graduates of the voc-tech program thought their education had helped them immeasurably to learn on the job. However, they did not have the same opinion about the college program. The journeymen thought that most of what was taught at the college was irrelevant to a tradesperson. Several of them argued that the college program was really a barrier to training apprentices, because apprentices spent too much time worrying about their college courses, which only distracted them from mastering their trade.

Even the most successful graduates of the new apprenticeship program agreed with the other journeymen that a considerable part of the college program was irrelevant. As with the lab techs, the theory courses were considered the most difficult and most irrelevant classes in the program. These courses were taught primarily to engineering students, and the instructors often assumed that everyone in the class was an aspiring engineer. Most of the apprentices' instructors were either engineers or scientists. None of the instructors or other students were skilled tradespeople. No one in the community college courses had ever worked in the trades. Several apprentices felt that the instructors condescended to them, once they learned that the apprentices were not engineering students.

African American production workers also expressed dismay about the college program. Although the skilled trades were largely a white male domain (4 percent African American and 1 percent female), African American workers thought minorities would have greater difficulty than whites with the college program. On average, the minority workforce had had less schooling, and had often attended the worst schools. Consequently, fewer minorities could qualify to bid for the apprentice openings. The African American workers believed that many minority workers who could do skilled tradeswork would be screened out by the college requirements. Since the new program had begun, only one African American worker had graduated, an individual who everyone agreed was of exceptional ability and tenacity. The minority workers also thought any fair evaluation of an apprentice should be based only on his or her job performance; grades in the school program were irrelevant. School was only considered useful when it supported workplace learning and achievement.

The skilled trades supervisors were also critical of the new program. They, too, found that the apprentices were preoccupied with their college courses. They added that the college failed to provide apprentices with training in some basic skills, such as blueprint reading, which had been taught in the voc-tech program.

According to the tradesworkers, the move to the college to improve the "quality" of journeymen implied that management believed that they were inferior tradespeople; a conclusion with which they disagreed and which made many of them angry. The union local president, a skilled trades journeyman, believed that the original resolution of the instrument repair training problem was a mistake. In every venue we observed the tradespeople voicing their objections to the college program. They fought vigorously to get the program moved back to the voc-tech and placed under the control of educators who were friendly to the tradesworkers' knowledge system.

Management, on the other hand, resisted the skilled trades' demand to relocate the course work. They pointed to the building trades, who had largely moved their apprenticeship programs out of the voc-tech program, as a positive example of the new system. The building trades, however, still retained control over their programs, whether they were located in their own facilities, in community colleges, or in voc-tech high schools. In addition, their programs had sufficiently large enrollments for them to hire their own instructors and exert influence over course content and teaching.

In a recent evaluation of the apprenticeship program, management had hired an industrial psychologist to validate the training that the apprentices received in college. It is likely that management assumed the consultant would support the new college-based program. However, the psychologist found that on average, two of the ten courses in each program could not be defended as job-relevant if challenged in an EEO complaint. As a result, courses were changed, and a blueprint-reading course at the voc-tech, along with one or two other courses, were added. In one respect, this change represented a union success; they had succeeded in influencing the nature of apprenticeship training for the trades. At best, this solution was a compromise, as both sides remained committed to their knowledge systems and their respective educational programs for training apprentices.

### *The Contest for Control over Knowledge Systems*

Lab techs had lost their limited control over their knowledge system when the apprenticeship program was abolished. The new associate's degree requirement not only deprived lab technicians of control over their formal education, it disrupted their workplace training system. The college programs, taught by scientists, emphasized theoretical training over practice-based learning. Technicians, however, learned diluted science. In the workplace, peer-trained task competencies had been replaced with a formal system which placed the responsibility for training in the hands of supervisors and scientists, who had never done lab technicians' work. Although some technicians, particularly in R&D, acquired a craft-type knowledge through their lab practice, many of them were not cognizant of their own knowledge system and, more importantly, it was not institutionalized. Instead, their knowledge was deprecated and, in R&D, subsumed within the larger scientific knowledge system. Since they lacked control over their own knowledge, lab tech work was largely shaped by the demands of others. In R&D it was shaped by the demands of scientific research, and in QC it was shaped by the logic of manufacturing.

Lab techs in both R&D and QC were supervised by those who had never done their work, but who held higher credentials. Consequently, their upward mobility was limited. Without further education they were restricted to moving to another technician position. While this might cause them to appear like professionals, the lab technicians were in fact the bottom of the scientific status and authority hierarchy.

In the QC labs, their work was integrated into pharmaceutical manufacturing and thus resembled factory work. They needed to conform to the highly interdependent demands of advanced manufacturing and information technologies. The least skilled work of the QC technician—measuring, counting, and weighing—had been automated. QC lab techs complained of highly routinized and boring tasks which were periodically punctuated with quality or machine errors. We found support for an observation, made over two decades ago, that the division of labor leads to boring and repetitive tasks for an increasing number of jobs performed by laboratory and quality-control technicians (Roberts et al. 1972).

In the R&D labs, technicians worked for scientists and they had to learn to be flexible, to adapt their practice to the scientists' idiosyncrasies, and to get along without complaining. Even those who took initiative and assumed extra responsibilities could only hope to "assist" a scientist in the development of new methods. Although they believed that they did all of the work, most of this work was unacknowledged, unappreciated, and invisible. They were nonprofessionals living within a professional knowledge system. However, their work was often more independent and less tightly controlled, and their practices more diverse and skilled, than those of their counterparts in the QC labs.

In contrast, the skilled tradesworkers were at the apex of the blue-collar status hierarchy. Although embattled with an engineering reconception of their occupation, they retained control over their shop floor knowledge system and maintained autonomy in their work practices. Nevertheless, they had lost some control over the formal education and socialization of apprentices. Yet, because their knowledge system was institutionalized, making them a cohesive social force, they had been able to mount a vigorous campaign through the union against the new educational system, gaining some unlikely allies. They were supervised by former tradesworkers, who were themselves products of the trades' knowledge system. The tradesworkers remained independent and largely defined the nature of their maintenance work, often working without any direct supervision. Journeymen were still responsible for the workplace education of apprentices. Apprentices and journeymen were also joined in a common cause to eliminate the college program and reclaim control over their craft knowledge system.

The differences between the skilled tradesworkers and the lab technicians suggest another dimension of dissonance for the lab techs. While both groups of workers were technically skilled, neither enjoyed professional prestige; however, their social experiences were markedly different. Tradesworkers knew their interests; they were organized, and they were consciously engaged in a struggle with management to maintain control over their knowledge system. They fundamentally appreciated that an autonomous, independent, skilled practice depended on their collective action. The lab technicians, however, were neither so cognizant nor so organized. Their identity as educated, technically competent employees with professional aspirations corresponded neither with the routine structure and nature of their work in the QC lab, nor with their inferior social status in R&D. Yet the lab technicians were not prepared, organizationally or ideologically, to take collective action to gain control. They also did not view the union as a means of support for such a struggle. As a result, their status was ambiguous and largely defined by what they were not. They did not possess professional prestige, autonomy, independence, authority or pay; however, they desired it. They also did not share craft organization, autonomy, or control, nor did they accept the production workers' subordination to managerial authority. Consequently, they persisted in a state of chronic personal dissatisfaction.

### *Lab Technicians on Management and the Union*

Technicians at this facility worked within either the R&D occupational hierarchy or the QC manufacturing bureaucracy; each division organized the work of technicians in markedly different ways. However, R&D and QC lab techs shared similar perceptions about themselves as technicians and their experiences with others. Many technicians, both QC and R&D, thought that they were not respected by supervisors, scientists, or upper management. Technicians felt that those in the upper echelons looked down upon them, saying, "They're not suppose[d] to socialize with us." Technicians reported, "We do all the work," and "Scientists develop new methods, and sometimes we get to 'assist' them." One lab tech related the following story: "A director wrote an official memo to upper management about how dumb and unmotivated we are. He referred to us as 'lemmings.' Do you know what that is? Well, I didn't, so I looked it up—it's a little brown rodent!" A copy of this memo circulated among the lab technicians, and was a source of considerable irritation.

Their consistently negative views toward management, however, did not translate into support for the union. Lab techs were highly critical of the union, telling us, "The union never defended us. They don't care about us," "The union is a puppet of management," "There is no union; there's only dues!" and complaining that the leaders "are all in the skilled trades." Many lab techs thought the union was run by and for the skilled trades, which was not completely accurate. The local leadership was a coalition government, drawn from two constituent groups: the skilled trades and the African American workers, who were primarily based in production. Both of these groups shared similarly hostile views about the changes in the skilled trades apprenticeship program; their leaders were surprised to learn that technicians had any problems.

## WORKERS' SELF-PERCEPTION OF SKILL AND JOB SATISFACTION

We developed the survey instrument in discussions with the Joint Labor-Management Apprenticeship Council. As the council sought consensus on the inclusion of each question, some important but politically sensitive items about supervisors, management, and the union were deleted from the final survey. As noted earlier, our survey results were based on the responses of 337 workers, including forty-three laboratory technicians, ninety-six skilled tradesworkers, and 198 semi-skilled (mainly production) workers. The respondents voluntarily filled out the written employee survey on company time.

Table 2.3 compares self-perceptions regarding the skilled tradespeople's, lab techs', and production workers' satisfaction, skill, and job security. It also reports

TABLE 2.3
Occupational comparisons (Five-point scale: 5 [Strongly Agree] to 1 [Strongly Disagree])

Occupational Comparisons of Self-Perception Variables and Scales

| | Occupations | | | | |
|---|---|---|---|---|---|
| | *Skilled trades* | *Are trades & lab techs' differences significant?* | *Laboratory technicians* | *Are lab techs' & production workers' differences significant?* | *Production workers* |
| Satisfaction | 3.92 | yes[c] | 2.97 | yes[a] | 3.34 |
| Skill | 4.21 | yes[c] | 3.68 | no | 3.46 |
| Autonomy | 4.16 | yes[c] | 3.07 | no | 3.14 |
| Complexity | 4.24 | yes[a] | 3.94 | yes[a] | 3.60 |
| Market value of skill in 5 years | 4.17 | yes[c] | 3.14 | yes[b] | 2.61 |
| Computer know-how | 2.74 | yes[c] | 2.98 | no | 2.82 |
| Job security | 3.51 | yes | 2.98 | no | 2.82 |
| Occupational comparisons of demographicvariables | | | | | |
| Female (1,0) | 0.02 | yes[c] | 0.44 | yes[c] | 0.69 |
| White (1,0) | 0.94 | yes[c] | 0.63 | yes[a] | 0.45 |
| Age (years) | 46.60 | yes[a] | 42.04 | yes[a] | 45.43 |
| Education (1–9) | 5.90 | yes[c] | 6.66 | yes[c] | 4.51 |
| Job tenure (years) | 14.86 | yes[b] | 10.38 | no | 8.15 |
| Seniority (years) | 20.39 | no | 16.95 | no | 15.80 |
| Married (1,0) | 0.85 | yes[c] | 0.56 | no | 0.56 |

*Note:* T-test significance levels for differences not zero are:
[a]$p < .05$, [b]$p < .001$, [c]$p < .0001$

demographic comparisons by occupation. We examined whether the differences reported by each occupational group were statistically significant using standard t-tests. Where the differences between the responses of skilled tradespeople and lab techs or between production workers and lab techs were significant, such is identified with a "yes," and the level of significance is also indicated. If the difference between the groups' responses was not statistically different from zero, a "no" was recorded.

Satisfaction, skill, complexity, and autonomy were measured on scales that we constructed using multiple items. Each scale's reliability coefficient (Cronbach's alpha) and respective items are reported in Table 2.4. Skill was measured as a two-dimensional construct, capturing both substantive complexity and autonomy (Spenner 1990). As in prior research, these two dimensions were highly correlated. When the ten items that measure complexity and autonomy were combined into a single skill scale, the scale had a higher degree of internal reliability than did either of the two subscales (alpha =.90, as compared to .86 and .71, respectively). Job satisfaction was measured by a four-item scale with a highly acceptable level of internal reliability (alpha = .80).

Returning to Table 2.3, lab techs were the most dissatisfied occupational group we surveyed. The self-report data indicated that most lab techs rated their jobs as having skill requirements similar to those of production workers. Nevertheless, their jobs were substantively more complex than the report the production workers gave of their jobs; yet lab tech jobs ranked the lowest on autonomy

TABLE 2.4
Satisfaction, skill, and education scales (Five-point scale: 5 [Strongly agree] to 1 [Strongly disagree])

Satisfaction (Cronbach's Alpha = .80)

- I am satisfied with the work I do at the Company.
- My work is interesting and challenging.
- I am satisfied with my opportunities for advancement.
- I intend to stay in my present job for as long as possible.

Skill (Cronbach's Alpha = .90) = Complexity + autonomy (10 item scale)

Complexity (Cronbach's Alpha = .86)
- My job requires me to use my memory.
- My job requires accuracy and precision.
- My job requires knowledge of math.
- My job requires reading and spelling.
- My job requires the ability to clearly communicate.
- My job requires eye and hand coordination.
- My job requires concentration.

Autonomy (Cronbach's Alpha = .71)
- I have the opportunity to exercise my own judgment on the job.
- My job requires me to regularly solve problems.
- I use all my skills from my experience and training on my present job.

Education

Please specify the highest level of education you have completed (circle number).

1. Some grade school
2. Completed grade school
3. Some high school
4. Completed high school
5. Vocational training school
6. Apprenticeship program
7. Some college (includes 2-year associate's degree)
8. Completed college
9. Some graduate work

(though not significantly different from the production workers.) In comparison with the production workers, lab techs thought that their skills would be more valuable in the market in five years, but they did not necessarily feel secure in their current jobs. The lab techs ranked significantly below the skilled tradesworkers in terms of job satisfaction, skill, and job security. On only one item, knowing how to use a computer, did the lab techs report greater competence than both the trades and semi-skilled workers.

The lab techs were also the youngest and best-educated group we surveyed. They were significantly more racially integrated than the skilled tradesworkers. Lab tech employment of women reflected the overall labor force's proportion of female employment, which was in sharp contrast to the skilled trades. On the other hand, the production workforce we surveyed was mainly nonwhite and female. All groups of workers exhibited long tenure in their current jobs, but the skilled trades had significantly more time in their jobs than did either the lab techs or the production workers. There was no significant difference, however, among the groups in company seniority. Skilled tradesworkers were significantly more likely to be married than either the lab techs or the semi-skilled workers.

When we examined the individual items that comprise the scales, we found much of the lab techs' dissatisfaction to be associated with their lack of autonomy, their routinized work, the dearth of formal opportunities to exercise judgment and solve problems, their inability to use their skills on the job, insufficient opportunities for advancement, and lack of acceptance by their co-workers. On each of these items, their attitudes more closely matched those of production workers and were significantly more negative than those held by the skilled tradesworkers. Yet in required job skills, such as math, reading, accuracy, and precision, their self-perceptions more closely approximated those of the skilled tradespeople rather than those of the production workers.

Table 2.5 reports the coefficients from three linear models that predict employee job satisfaction, estimated using a population weighted least squares. Model 1 reveals the importance of skill in predicting satisfaction. Once skill and the future market value of skill were held constant, the difference between production workers (the omitted category) and the skilled tradespeople in job satisfaction became statistically insignificant, suggesting that the skilled tradespeople were production workers with more skill. On the other hand, lab techs remained significantly less satisfied. Even when we control for perceptions of job security, computer use, and demographic differences in models 2 and 3, lab techs remained significantly less satisfied than either the production workers or the skilled tradespeople. These results suggest that we need to look at factors other than skill, job security, or differences in worker backgrounds to explain the relative dissatisfaction of the lab techs.

TABLE 2.5
Predictors of job satisfaction

| *Dependent variable: Added satisfaction scale Mean = 13.24* | *Model 1* | *Model 2* | *Model 3* |
|---|---|---|---|
| Skill | 0.22[c] | 0.18[c] | 0.19[c] |
| | (.02) | (.03) | (.03) |
| Market value of skill | 1.02[c] | 0.65[c] | 0.61[c] |
| | (.15) | (.17) | (.17) |
| Job security | | 0.72[c] | 0.70[c] |
| | | (.15) | (.16) |
| Female | | 0.96[b] | 0.92[b] |
| | | (.37) | (.38) |
| Seniority | | 0.04[a] | 0.03 |
| | | (.02) | (.02) |
| Job tenure | | | −0.01 |
| | | | (.02) |
| Computer | | | −0.19 |
| | | | (.16) |
| White | | | −0.12 |
| | | | (.37) |
| Married | | | −0.21 |
| | | | (.36) |
| Education | | | −0.12 |
| | | | (.13) |
| Skilled trades | −0.65 | .26 | 0.61 |
| | (.51) | (.56) | (.65) |
| Lab technician | −2.15[c] | −1.79[c] | −1.34[b] |
| | (.43) | (.43) | (.53) |
| Intercept | 2.89[b] | 1.89[a] | 3.16[b] |
| | (.81) | (.84) | (1.13) |
| Adjusted R-squared | .44 | .50 | .48 |

*Note:* T-test significance levels for beta not zero are:
[a]$p < .05$, [b]$p < .01$, [c]$p < .0001$
F values for each model significant at $p < .0001$

## *Can Technicians Construct an Independent Identity?*

In this section we attempt to clarify the social portrait of technicians. The conclusions we reach are tentative, and we hope they will serve to stimulate further research on this important occupational group and their position in the technical division of labor. The *Current Population Survey* data indicated a high level of association, interdependence, and interconnection between technicians and technical professionals in several dimensions, including employment and earnings growth, unionization, and the gender composition of employment; most of the detailed occupational categories also reflected these same close links. We interpreted the lagging growth patterns of technicians in employment and earnings as evidence that technicians serve mainly to complement technical professionals in the technical division of labor. This explanation also suggests that the demand for technicians' labor is derived from the demand for technical professionals. These propositions can be more formally tested using standard econometric techniques to examine labor-labor substitution or complementarity of technicians and technical professionals. Our other conclusions, however, will require more qualitative or less standard quantitative evaluations. They also center on the interdependence of technicians and technical professionals.

We observed a widening earnings gap between technical professionals and technicians. Technicians earned 72 percent of the median professional weekly earnings in 1983, but by 1993 the gap had expanded, and technicians earned only 65 percent of professional earnings. However, if technicians primarily complement professionals in the technical division of labor, as we believe they do, then why has the real wage growth of technicians lagged so far behind the substantial increases gained by technical professionals during the last decade? The answer, we suspect, resides in the technical division of labor. Each profession claims all knowledge and practice within its exclusive domain (Abbott 1988). Because technicians work within the domains of professions, their contributions are largely invisible. Consequently, not only are technicians' contributions unrecognized, but the rewards based on those contributions, we suspect, are largely appropriated by the highly visible technical professionals. This would partly explain the widely reported feelings of relative deprivation held by technicians, the pay dissatisfaction expressed by the surveyed engineering technicians, and the contrasting resentful statements of the lab techs that "we do all the work," while "sometimes we get to assist" the scientists. It is also worth noting that the more highly educated lab techs earned less than the skilled tradesworkers and little more than some production workers. These disparities are engendered by the technicians' relationship to professionals in the technical division of labor.

Technicians, in all aspects of their occupational life, are formally subordinate to the technical professions. In our case analysis, regardless of whether lab technicians worked in an organizational or occupational division of labor, they were supervised by professional scientists. Technicians were formally excluded from supervision because they lacked sufficient educational credentials to qualify for consideration, i.e., they were not professionals. Consequently, technicians were always supervised by people who had never performed their jobs. Also, in the 1970s, as education and training requirements for technicians were changed and formalized, lab technicians were denied any formal role in the indoctrination of new technicians. In the educational process, the community college programs relied exclusively on technical professionals for educating student technicians. These professionals often communicated professional standards and served as professional role models for the socially aspiring technicians-in-training. In the workplace, the apprenticeship model, with its task competencies, which provided an explicit role for the experienced technician in the training of technician apprentices, was replaced by a formal training program under the direction of the supervisor-professional scientist. Consequently, regardless of how skilled their practice may be, technicians do not control their occupational knowledge or skill system, nor do they control the entry, education, or formal training of new recruits. In this sense, *they are craftworkers without craft control.*

Our comparison of the lab technicians with the skilled tradesworkers, we believe, illustrates precisely why technicians are not new craftspeople (at least not yet). In all dimensions of their occupational roles, technicians are largely defined by technical professionals. Technicians, however, often believe they possess knowledge and skill superior to that of their professional colleagues, but rank themselves significantly lower on prestige, opportunity for advancement, pay, and authority (Evan 1964; Koch 1977; Roberts et al. 1972). This higher self-appraisal of their technical performance skills, coupled with their lower status, reportedly leads to a sense of relative deprivation, which has been the alleged source of technicians' dissatisfaction. Yet, a comparison with the skilled trades suggests there must be more to this story.

In contrast to the relatively passive history of technicians' occupations, the skilled trades have been engaged in a jurisdictional battle with the engineering profession since the rise of scientific management a century ago. Skilled tradesworkers, who on virtually all skilled practice dimensions possess superior manual abilities when compared to engineers, manifested none of the feelings that reflect status inconsistencies, relative deprivations, or simple dissatisfaction. Instead, the skilled tradespeople judged their competence based on the knowledge and practice standards of their respective trades; their relative standing in relationship to engineering was unimportant to them. When questioned, they explicitly eschewed any professional aspirations; they enjoyed being tradespeople. The trades also vigorously resisted any redefinition of their standards by the engineering profession, as witnessed in the dispute over the community college educational program. Only when the former apprentices reported feeling that they were condescended to by their college instructors did we detect any insecurity in their reactions.

Most notably missing in technicians' occupations is any formal arrangement permitting technicians to set entry and performance standards for their occupation, or allowing them to train, educate, or socialize new recruits or supervise other technicians. Instead, technical professionals, not technicians, play all these crucial roles. As a result, professionals largely control the social definition of technicians' occupations. Technicians, consequently, are subordinated within a division of labor, regardless of whether it is organizational or occupational, without any real institutionalized power over their occupation or work. We believe that it is this lack of any formal institutionalized power over their occupations that is the main source of their widely reported dissatisfaction. It also serves to explain much of the heterogeneity of technicians' occupations that has been reported in the research literature.

We found no evidence to suggest the emergence of a new class system (Bell 1973). Rather than class, we suggest that there is a process of increasingly formal stratification in the technical division of labor which places technicians at the bottom of the technical hierarchy without any formal control over their work lives. In this sense, they are not unlike factory operatives in the industrial division of labor. This lack of decision making authority may legitimate why technicians earn basically the same wages as skilled blue-collar workers, even though they possess considerably more education.

The Current Population Survey data on both real wage growth and qualifying education, however, placed technicians as an intermediate occupational category in between technical professionals and skilled blue-collar workers. The growing identification of technicians with two-year associate's degrees suggests a possible institutionalization of this unique intermediate position, making them neither craftworkers nor professionals. While these objective changes suggest an increasingly important role for technicians, ultimately the fate of these occupations will be determined by the technicians' subjective understanding of their occupational role and whether they can forge a unique and independent identity.

We believe that a key element in forming a clear identity is self-organization. Whether they adopt the professions' associational model or the skilled trades' craft union model, eventually technicians must organize themselves if they want to gain control over their knowledge and skill system. This means institutionalizing their education, training, socialization, and supervision under their own control, while securing formal autonomy and independence over those practices in which they are expert. Without a doubt, such an undertaking would directly challenge the status of the respective governing profession. Consequently, if it is possible for history to foretell the future, technicians can expect no help from the professions in establishing an independent identity. Instead, technicians will be faced with a myriad of reasons why they cannot organize. Where these arguments fail and organization is viewed favorably, following in the footsteps of company unions, the professions may establish professional-dominated technicians' associations.

Technicians, as a group, are skilled, educated, and technically competent. This gives them the potential for exercising power in the workplace. More important, technical professionals have grown highly dependent on technicians. The greatest barriers technicians face in their own occupational organization are ideological. Given that widely held individualism did not win independence or autonomy for either the professions or the crafts, collective action based on solidarity may be necessary. Technicians face an obvious choice to either persist in a state of chronic personal dissatisfaction or to take charge of their occupational futures. Researchers will need to pay special attention to all technician efforts to organize, associate, or unionize, since such efforts may greatly determine the occupational division of labor and social equity in postindustrial society.

# 3

# WHOSE SIDE ARE THEY ON? TECHNICAL WORKERS AND MANAGEMENT IDEOLOGY

*Sean Creighton and Randy Hodson*

## INTRODUCTION

There has long been debate concerning the position of technical workers in the class structure (Mallet 1975; Gorz 1967, 1972; Whalley 1991a). This debate is of interest because it identifies technical workers as a growing part of the workforce who occupy an intermediate position between management and labor. Three contrasting pictures of technical workers emerge from this debate. The first image is of technical workers as a new and dynamic part of the working class, capable of regaining the kind of power that skilled craftworkers once possessed in the production process. The second image is of technical workers as part of a newly emerging middle class whose education and control over scientific knowledge lead them to identify with management. A third view, based on organizational case studies of engineering firms, argues that technical workers, especially engineers, differ from both blue-collar workers and management in terms of their work practices. This research develops the theme that engineers are "trusted workers" tied to management through a series of perquisites, such as promotion opportunities, which are not available to other non-managerial workers (Whalley 1986b).

Serge Mallet argues that what defines the position of the worker in the modern economy is the loss of occupational autonomy as well as the loss of economic initiative. He points out that, over time, workers have shifted their demands from issues of control over what work is performed and how that work is to be carried out (the labor process) to demands relating to their standard of living and position as consumers (Mallet 1975, 22).

The growth of technical work, however, is a crucial development because as research, invention, and quality control become more important in the organization of production, technical workers, who are in charge of these new innovations, potentially regain on a collective level the craft autonomy that workers possessed prior to the Industrial Revolution and the development of modern systems of mass production (Mallet 1975, 23). André Gorz wrote that technical workers approach work in a manner:

> [. . .] which measures the scientific and technical potential of an enterprise in scientific and technical terms and which sees this 'technological capital,' this 'human capital'—the cooperation of polished teamwork, the possibility of conquering new domains of knowledge, new chances for the domination of man over nature—destroyed by the barbaric command for financial profit (Gorz 1967, 7).

Thus, the perspective of technical workers on knowledge, teamwork, and the illogic of capitalist relations leads to a growing identification of their common interests as workers. Gorz argued that technical workers find that creative work, love of workmanship, and longrange research goals are incompatible with the criterion of profitability.

According to this viewpoint, the position of this new working class in the production system leads it to see the deficiencies inherent in the organization of modern capitalism and to arrive at a consciousness of a new way of organizing productive relationships. Mallet claims that the demands of the new working class for control of the process of production will call into question the entire hierarchical nature of industry. It is "the producers of sciences and techniques, more than [of] products, who will build the true industrial society, purged of its capitalist and technocratic archaisms" (Mallet 1975, 32). In other words, Mallet proposed that because of their skill, autonomy, and position in the production process, technical workers will develop an oppositional consciousness toward management.

While these concerns may appear to belong to another era, they foreshadow modern concerns about the relationship between technical workers and management in an increasingly competitive international environment. There is no doubt that management in modern high technology firms experiences some of the same contradictions described by Gorz and Mallet. Management's efforts to control technical workers are organizational and cultural responses to the demands of modern production: high quality, innovation, flexibility, and profitability (Kunda 1992). The role and orientation of technical workers is central to these issues. While some of these questions have been resolved in relation to engineers, in this paper we focus on the position of the newly emerging occupation of technicians.

In this chapter, we evaluate Mallet's hypothesis concerning the role of technicians in advanced industrial society. Our lead questions will be: "What are the objective characteristics of technicians in terms of education, skill, autonomy, and related attributes?" and "Do technicians subscribe to or contest management ideology?"

## TECHNICAL WORKERS: WORKING CLASS OR MIDDLE CLASS?

Not everyone shares Mallet's view of the nature of the new working class. There has been much debate about the location of technical workers in the class structure. Alain Touraine (1971) assigned this group to a "new middle-class strata" because, although they display elements of the labor process that tie them to the working class (craft knowledge, lack of control over their work), they also show characteristics that objectively link them to management (formal education, belief in the value of technological progress, tendencies to act individually rather than collectively). Herbert Marcuse (1968, 280) also casts doubt on the idea that white-collar workers in general will develop an independent political consciousness. Similarly, the work of Harry Braverman (1974) gives no indication that the technical segment of the working class will rise up against management or that these workers will be able to counteract effectively the processes that tend to reduce their control of their own labor.

By the 1970s, Gorz had changed his view of the nature of the new working class. His theoretical stance began to take into account both the technical and the ideological position of technical workers. He eventually came to have quite a negative view of the likelihood of technical workers taking a leadership role in transforming workplace relations:

> Whether we like it or not, we must see technicians in the manufacturing industries as key instruments of the hierarchical regimentation required by the capitalist division of labor; their role is to make sure thereby that the maximum labor and surplus value is extracted from each worker. Their role is to de-qualify workers by monopolizing the technical and intellectual skills required by the work process. They embody the dichotomy between manual and intellectual work, thought and execution. They hold significant financial, social and cultural privileges. They are the workers' most immediate enemy: they represent the skill, knowledge and virtual power of which the workers have been robbed. In a machine tool shop, every technician hired will turn five, ten or twenty hitherto skilled workers into unskilled underdogs, thereby enabling the boss to pay them unskilled wage rates. (Gorz 1972, 34)

Gorz (1982) argues that the technicians' superior position in the workplace does not come from a monopoly on useful knowledge, but from a monopoly on useless knowledge. As Keefe and Potosky (Chapter 2) point out, there is often a requirement for technical workers to pass abstract science courses that the workers themselves claim are irrelevant to their workplace practice. As they and others have pointed out, the function of such knowledge certification goes far beyond its content (Keefe and Potosky, Chapter 2; Scarselletta, Chapter 8; and Bucciarelli and Kuhn, Chapter 9). The hierarchically superior position of the technical worker comes from culturally validated but abstract knowledge acquired in the school system (cf. Koch 1977; Hodson, Hooks, and Rieble 1992). Gorz argues that, from a political viewpoint, there is an unbridgeable objective class distinction between technical staff with supervisory functions and production workers.

According to this perspective on technical workers, there is very little chance that technical workers could be brought into alliance with the working class. During normal times, technical workers cannot be expected to support the demands of their fellow workers because:

> [. . .] the work of technological application of science takes place under the sign of the dominant ideology, which they materialize even in their "scientific" work; they are thus supporters of the reproduction of ideological relations actually within the process of material production. Their role in this reproduction, by way of the technological application of science, takes the particular form under capitalism of a division between mental and manual labor, which expresses the ideological conditions of the capitalist process. (Poulantzas 1975, 236–7)

Gorz (1972, 37) points out that when technical workers do rebel, it is frequently not in solidarity with the working class, but against being treated as proletarians. He argues that where technical workers have rebelled, it has been against the hierarchical division, fragmentation, and meaninglessness of their work, or against the loss of all or part of their social privileges. He regards their actions as attempts to regain middle-class status rather than as precursors to a leadership role in a new unified working-class movement. The privileges of engineers and technical workers are inherently tied to the capitalist order, and they act as direct agents of capital. Technical workers function as intermediaries between capital and labor, supposedly possessing a monopoly on technical knowledge. Through this monopoly they act to ensure the subjugation of the rest of the working class. In this vision, technical workers are seen as the recipients and keepers of knowledge of production processes originally taken from craftworkers (Braverman 1974).[1]

1. It should be noted that this vision of the role of technical workers is heavily influenced by studies of engineers in industrial settings. However, technical workers frequently operate in environments where there may not be any lower-level workers. Thus, whether these insights may be generalized to all technical work is debatable.

In the writings of neo-Weberians on the "new middle class," we find a set of propositions which agree with the later Gorz and directly contradict the assertions of Mallet and the early Gorz (Giddens 1982, 40–41; Goldthorpe 1982, 183; see also Wuthnow 1986; Lee and Smith 1992). From this viewpoint, technical workers base their claim to status and power on their occupational roles, not their ownership of the means of production, and thus belong to a "new middle class." Due to their education and control of scientific knowledge, their identification tends to be with the technocratic elite. They form an occupational status group that is normatively integrated within the enterprises in which they work and within middle-class society at large. The claim of "working-class consciousness" for technical workers is inconsistent with these arguments that technical workers differ from the traditional proletariat in terms of their skills, market situation, work situation, status, and identity.

Whalley (1986b) uses a comparative case study approach to examine the structural position of technical workers in the labor process and their social and ideological position relative to other workers. Whalley (1986a, 1991a) argues that engineers belong to the category of trusted workers. These workers carry out strategic jobs within the corporation and are granted a significant degree of autonomy to do so. A social schism divides trusted workers from wage workers. The boundary between trusted workers and wage workers marks the transition between different employment contracts. The "loyalty package" that employers offer to attract and maintain a staff of trustworthy incumbents for technical-supervisory positions often includes higher salaries, fringe benefits, job security, more vacation time, more flexible working hours, and, in many firms, such symbolic distinctions as separate parking lots and dining areas. This contrasts with the cash nexus that binds hourly wage workers to employers. John Westergaard (1984) similarly describes technical workers as belonging to the world of "careers" rather than the world of "jobs."

Whalley's image of engineers as trusted workers rests on three propositions. First, management avoids the proletarianization of technical work due to its need for "trusted" employees. It is not in the employer's interest to de-skill the pool of workers from which higher management will be drawn. Second, managers and engineers should not analytically be placed in different classes—they are routinely part of the same career line linked by promotion and career trajectories. Finally, technical knowledge only becomes expertise when it is accepted as such by employers—craftspeople and other workers continue to possess much essential knowledge about the production process, but their knowledge is not sanctioned as such by employers. Engineers' knowledge, in contrast, is sanctioned as expertise through the process of granting them trust and responsibility. Gideon Kunda's (1992) analysis of a large high-technology engineering firm implicitly supports this argument. He observes the resistance of the firm to downsizing its engineering workforce during recession, the routine integration of technical workers into the management hierarchy, and the embeddedness of engineers in an organizational culture, which puts the onus on them to "do what's right" for the organization (Kunda 1992, 71–72).

Whalley (1986b) maintains that what divides engineering workers from other workers is a hierarchy of discretion, not of knowledge. In contrast to other theorists, he claims that it is not formal knowledge that determines the position of technical workers in the hierarchical division of labor. What is important from the management perspective is the responsibility to use expertise appropriately in an environment where technical knowledge is only a small part of the skills needed to accomplish management's prescribed agenda. Job security, promotion opportunities, and autonomy are rewards for those who occupy positions seen by management as important to the profitability of the firm. Many of the tasks that have been taken away from skilled craftspeople and other workers are now carried out not by management but by engineers tied to management by promotion prospects, material privileges, and ideological constructs (see also Ng 1986)[2].

As Peter Whalley and Stephen Barley (Chapter 1) point out, engineers are the "prototypical technical workers of high industrialization," and the anomalies implicit in their labor and social location have played havoc with both occupational and organizational understandings of work. The understanding of engineering as a "corporate profession" tied to management leads us to ask if technicians are also incorporated into occupational and organizational structures as trusted workers. As Jeffrey Keefe and Denise Potosky (Chapter 2) point out, the growth of technicians' occupations seems to be tied to the growth of technical professionals. However, conditions at the plant they studied do not seem to suggest that technicians enjoy the same control of the labor process or the same rewards as professionals. The propositions underlying the trusted worker thesis, developed to understand the position of engineers, allow us to generate a series of hypotheses related to technicians that should help to shed light on these issues.

Our analysis of the issues that flow from this discussion will proceed in two parts: First, we will profile technicians and contrast their profile to that of other occupational categories. Second, we will examine the adherence of technicians to management ideology.

2. Whalley's approach may seem peculiarly suited to the British case, because in Britain, many engineers are promoted from the shop floor and the role of formal qualifications has not been historically strong. However, his theoretical argument may be valid cross-nationally. What engineers may be ultimately rewarded for is their willingness to "do what's right" for the company regardless of the source of their technical skill (i.e. university degree versus apprenticeship).

### *Hypotheses about Technical Workers and Management Ideology*

Technicians have been described as having more education, greater occupational prestige, and more on-the-job training than other workers. At the descriptive level, we expect to find that technical workers are less well educated than professionals but better educated than other occupational groups. Barley (1991, 15) states that "with the exception of professionals, technical workers constitute the most highly educated occupational category." He also suggests that they will have somewhat lower occupational status than professionals. Many authors have suggested that on-the-job training is a key characteristic of technical work (Zussman 1985; Orr 1996). "Members of such [technical] occupations may require considerable formal knowledge of science, math and technology, yet their most valued skills appear to be those developed in a hands on conversation with materials and techniques" (Barley 1991, 20; see also Whalley 1986b, 54–8). Thus we would expect technical workers to have extended training time on the job. Therefore, we propose:

> *Hypothesis 1*: Technical workers will have higher levels of education, occupational prestige and time required to learn their jobs than other workers.

Whalley has argued that engineers are tied to management by a loyalty package of benefits not available to other workers. If the trusted worker thesis is applied to technicians, it leads us to predict:

> *Hypothesis 2*: Participation, promotion opportunities, and job security will be higher for technical workers than for other workers.

It has been argued that technical workers' monopoly of technical knowledge is necessary to ensure the subjugation of the rest of the labor force. Technical workers are seen as the recipients of production knowledge taken from craftworkers (Whalley 1986b, 9). Others have argued, however, that technical workers are not more skilled than other workers and that claims in this regard rest on officially sanctioned but empty credentials (Gorz 1982). To evaluate these claims, we set forth:

> *Hypothesis 3*: Technical workers will be more skilled than other workers.

Some scholars have attempted to explain the ideology of technical workers as a function of their class location as autonomous workers. Erik Wright (1978) recognized that, in the Anglo-American tradition, engineers often function solely as technical experts and are not responsible for the supervision of other workers. Thus, they do not necessarily directly perform the central function of management (supervision), as is often assumed by continental European writers (see Carchedi 1977). Wright thus distinguishes between management positions and expert positions (1978). This distinction appears to be even stronger when we examine the work of technicians (see Zabusky, Chapter 6). Technical experts are seen as occupying a contradictory class location in this scheme. Their autonomy is rooted in their control of the labor process (product design and how their work is done) rather than in their position in organizational hierarchies. Therefore, we predict:

*Hypothesis 4*: Technical workers will have higher levels of autonomy than other workers.

Specific hypotheses about the ideological orientation of technicians provide the most direct test of whether they fit the image of trusted workers. These hypotheses refer to worker consciousness and to the role of promotion opportunities, skill, and autonomy in determining consciousness. Engineers and technical workers are said to be tied to the capitalist order as the direct agents of capital. Whalley (1986b, 38) states that, despite salaries which are poor when compared with those of managers, engineers show little sign of worker consciousness. Therefore, we hypothesize that:

*Hypothesis 5*: Participation, promotion opportunities and job security will increase adherence to management ideology.

*Hypothesis 6*: Skill and autonomy will increase adherence to management ideology.

*Hypothesis 7*: Technical workers will have a higher level of adherence to management ideology than other workers.

In the following analysis we use gender and unionization as control variables because these have been identified as key determinants of oppositional consciousness (see Cavendish 1982; Fantasia 1988; Westwood 1984; Yount 1991).

## DATA AND METHODS

The data used to test the above hypotheses are derived from a telephone survey, conducted in the spring of 1992, of a random sample of employed people in a Midwestern state. The survey design was based on the random selection of telephone exchanges throughout the state and included cities, small towns, and rural areas. The telephone numbers called were generated through computer-assisted random digit dialing. Screening was conducted at each household contacted to determine whether any household member was at least eighteen years of age and currently working for pay. (Individuals who were self-employed or who worked without pay in family farms or businesses were excluded.) If more than one adult member of the household was working for pay, the appropriate respondent was computer-selected. Interviewing was regularly monitored by supervisors, with each interviewer monitored at least twice during each four-hour shift. The response rate for the survey was 70.0%.

### *Occupational Groups*

Employees are divided into four occupational groups. The first group is technical workers, which includes all census occupations listed as technicians and related support occupations (U.S. Census 1982). The second group is professionals, which includes doctors, nurses, lawyers, teachers, social scientists, social workers, counselors and writers. The third group is managerial and supervisory personnel. This group includes executives, administrators, and supervisors, regardless of specific occupation or industry. The final group is nontechnical workers, which includes all other occupations (mainly clerical, manual, and service occupations). In order to avoid confusion, we will refer to this group as "nontechnical workers" or "other workers." Where we discuss professionals, managers, and other workers simultaneously, we will refer to them as "other occupational groups" or "all other occupational groups."

We completely excluded engineers from this analysis for a number of reasons. First, as the "prototypical technical workers," sharing many of the same skills and work practices as technicians, it did not seem very appealing to group them with professional workers in contrast to technicians. Second, to have included them as technical workers in a category with technicians would have potentially muddied our evaluation of the social location and adherence to management ideology of technicians. Third, the attractive solution of creating a separate category for engineers was precluded by the fact that we had too few engineers in our sample to evaluate them separately (n = 8).

### *Education, Occupational Prestige, and Time Needed to Learn the Job*

Education is measured by the question: "What is the highest grade of school or level of education you have completed?" Answers to this question are coded as years of education completed. Occupational prestige is measured using a socioeconomic index adjusted for gender differences in perceived occupational prestige (Stevens and Cho 1985). Time needed to learn the job is measured by the question: "How long does it take a person to learn your job as a whole, in all aspects, not just the machinery?" The context of this question clearly indicated that the question referred to time on-the-job needed to learn the tasks. Responses were initially entered in whatever time frame the respondent used for his or her reply (e.g., "one week," or "six years") and were later recoded to reflect the number of days needed to learn the job.

### *Adherence to Management Ideology*

Management ideology is measured by the following item: "If given the chance, the non-management employees at my workplace could run things effectively without management." Respondents were asked to agree or disagree (on a four-point scale) with the above statement. Responses were scaled so that high values reflect disagreement with the above statement; high values thus indicate a belief in the necessity of management.

### *Trusted Worker Characteristics*

Participation is measured in two aspects: being consulted by one's supervisor and participating in organizational decisions. These aspects of participation are measured by level of agreement with the following items: "My supervisor frequently asks for my opinion on work related matters," and "My job allows me to participate in important decisions that concern my organization." The two participation items scaled with an alpha of .73. Promotional prospects are measured by the item: "Would you say the opportunities for promotion from your job are: excellent, good, fair, or poor?" Job security is measured by the following item: "If your job were to be phased out because of new technology or changing markets, which of the following situations do you think would happen?" Responses were coded as one if the respondent selected the response, "Your employer would find something else for you" (indicating job security). Responses were coded as zero (indicating job vulnerability) if the respondent selected either the statement, "You would be out of a job," or "Do you think something else would happen? And if so, what?" In every case in which respondents selected the latter option, they elaborated on this response to say that they would have to get a new job.

### *Labor Process Variables*

Skill level is measured utilizing responses to the following items: "I often get to learn new things on my job," "My job requires a high level of skill," and "My job allows me to use my full abilities." Respondents either agreed or disagreed (on a four-point scale) with these items. When the responses to these items are factor analyzed, a one factor solution emerges with an eigenvalue of 1.82. These three items scale with an alpha of .68. Autonomy is measured by level of agreement with the following question: "I can set priorities about the order in which I do required tasks," measured on a four-point scale.

### *Competing Statuses*

Gender is coded as 1 = male, 0 = female. Union membership is also measured as a binary variable and is based on the question: "Are you a member of a union?" (1 = yes, 0 = no).

## RESULTS

Average levels of education, occupational prestige, and time needed to learn the job for each occupational category are reported in Table 3.1. The results largely confirm prior expectations. Professionals have the equivalent of a college degree; technical workers and managers have the equivalent of an associate's degree or a little higher. Nontechnical workers report having, on average, the equivalent of a high school degree. Professionals have the highest levels of education, occupational prestige, and learning time while nontechnical workers have the lowest levels on all three characteristics.

Tests of statistical significance reveal that technical workers have significantly higher levels of all three characteristics—education, occupational prestige, and job learning time—than do nontechnical workers. It is also interesting to note that technicians are quite distinct from the professional group. They have significantly less education and occupational prestige than do professionals. They also have approximately one less year of on-the-job training, although this result is not statistically significant. Technicians and managers have similar levels of education and occupational prestige, despite the inclusion of many lower-level managers and front-line supervisors in the managerial category. In terms of time needed to learn one's job, managers require approximately one year more than technicians. The expectation that technical workers require even more on-the-job learning time than professionals or managers is not supported by these findings (see Smith 1987; Barley 1988a; Zussman 1985). In summary, technical workers have significantly greater education, on-the-job learning, and status than

TABLE 3.1
Education, occupational prestige, and time needed to learn job by occupational group (N = 363)

| | *Managers* | *Professionals* | *Technical Workers* | *Workers* | *Total* |
|---|---|---|---|---|---|
| Education (years) | 14.97 | 16.31[a] | 14.68 | 12.35[a] | 13.61 |
| Occupational prestige | 46.43 | 64.54[a] | 50.04 | 25.96[a] | 37.40 |
| Learning time (days) | 804.36[b] | 826.01 | 518.00 | 288.46[a] | 484.75 |
| N | 76 | 54 | 19 | 214 | 363 |

*Note:* Each group mean is tested against that for technical workers. Significance levels are denoted by a = .01, b = .05 (2-tailed t-test).

do nontechnical workers. They have less on-the-job training than managers and professionals and significantly less education and prestige than professionals.

Table 3.2 reports the means of the variables hypothesized to be important in explaining adherence to management ideology, evaluated separately for each occupational group. The trusted worker thesis leads us to expect that technicians will be integrated into the firm in a manner similar to managers, but the results presented in Table 3.2 do not strongly support this hypothesis. If the trusted worker hypothesis were fully borne out, we would expect no differences to emerge between the scores of technical workers and managers. Indeed, we find that no differences emerge between technical workers and managers in terms of promotions, job security, or having their opinions requested by supervisors. However, this evidence cannot be interpreted as strong support for the trusted worker thesis because technical workers are also not significantly different from nontechnical workers on these measures. It is clear from the data that technicians have less participation in the affairs of their companies than do managers, which casts further doubt on the trusted worker thesis.

The notion that technicians occupy a special position as trusted workers in an organizational space located between management and workers (but closer to management) is not borne out. Technical workers' levels of participation are not significantly higher than those of other workers. Their levels of perceived promotion opportunities and job security are broadly similar to those of other occupational groups and they participate in company decisions much less than do managers.

**TABLE 3.2**
Labor process, trusted worker, and competing status variables by occupational group (N = 363)

| | *Managers* | *Professionals* | *Technical Workers* | *Workers* | *Total* |
|---|---|---|---|---|---|
| Trusted worker variable | | | | | |
| Ask opinion | 3.54 | 3.41 | 3.21 | 2.90 | 3.12 |
| Participate in decisions | 3.43[a] | 3.24 | 2.84 | 2.60 | 2.89 |
| Promotion opportunities | 2.37 | 2.08 | 2.21 | 2.02 | 2.12 |
| Job security | .60 | .40 | .42 | .53 | .52 |
| Labor process variables | | | | | |
| Learn new things | 3.59 | 3.61 | 3.37 | 3.18 | 3.34 |
| Use full abilities | 3.34 | 3.46 | 3.42 | 2.88[b] | 3.09 |
| High skill | 3.41[b] | 3.80 | 3.74 | 2.79[a] | 3.12 |
| Autonomy | 3.79[b] | 3.78 | 3.42 | 3.13 | 3.38 |
| Competing statuses | | | | | |
| Male | 67%[b] | 37% | 37% | 49% | 50% |
| Union member | 1% | 18% | 00% | 28% | 19% |
| N | 76 | 54 | 19 | 214 | 363 |

*Note:* Each group mean is tested against that for technical workers. Significance levels are denoted by a = .01, b = .05 (2-tailed t-test).

Comparing occupational differences in the three indicators of skill, we see that technical workers are significantly more likely than other workers to say that their jobs require them to use their full abilities. Technical workers also evaluate their jobs as requiring a higher level of skill than either managers or other workers, though they have significantly less autonomy than do managers. Technical workers thus perceive themselves as having more skill than managers and workers, and skill equal to professionals, but this does not translate into professional or managerial levels of autonomy. This finding suggests the interesting possibility that, despite their skill, technical workers gain neither the participation that comes from a management position nor the autonomy that accrues to both managers and professionals.

There are significantly fewer women in the managerial category than in the technical worker or professional categories. The professional category includes a large proportion of female teachers and the technical category includes a large proportion of female health technicians. Management is the most male-dominated of the four occupational categories.

None of the technical workers are union members, whereas 28 percent of workers are unionized. In addition, 18 percent of the professional category are unionized. This latter finding is mainly due to unionization of professionals in such occupations as teaching. As we would expect, practically no managers or supervisors are union members.

Table 3.3 presents coefficients from the regression of management ideology on the occupational group, labor process, trusted worker, and competing status variables specified above. Model 1 indicates that, as expected, managers strongly endorse management ideology. Professionals also endorse the necessity of management. Technical workers are not significantly more likely than other workers to endorse management ideology.

Model 2 adds the trusted worker, labor process, and competing status variables to the equation predicting adherence to management ideology. The coefficients for managers and professionals remain significant and positive. The coefficient for technical workers decreases in size and remains nonsignificant. None of the trusted worker indicators has a significant influence on management ideology.[3] In summary, the possibility that technicians can be considered trusted workers receives little support from these findings.

The labor process variables do not predict adherence to management ideology. Neither having greater autonomy nor possessing higher skill levels significantly predicts increased adherence to management ideology.

3. As discussed in the Data and Methods section, "Ask opinion," and "Participate in important decisions," are combined into a summary participation scale. "Learn new things," "Use full abilities," and "High skill," are similarly combined into a summary skill scale. These sets of variables are scaled to avoid multicollinearity among these closely related indicators when they are entered into the multivariate analysis.

TABLE 3.3
Regression of management ideology on occupational group, labor process, trusted worker, and competing status variables (N = 363.)

| *Independent Variables* | *1* | *2* |
|---|---|---|
| Occupational group | | |
| Manager | .579[a] | .331[b] |
| Professional | .406[b] | .315[b] |
| Technical | .193 | .002 |
| Trusted worker | | |
| Participation | | −.041 |
| Promotion | | .099 |
| Job security | | .072 |
| Labor process | | |
| Skill | | .039 |
| Autonomy | | .042 |
| Competing ideologies | | |
| Male | | .205[b] |
| Union member | | −.515[a] |
| Constant | 2.159 | 2.046 |
| Adjusted $R^2$ | .045[a] | .079[a] |

*Note:* Unstandardized regression coefficients are reported. Significance levels are denoted by a = .01, b = .05 (1-tailed t-test). Nontechnical workers serve as the reference category for estimating the effects of occupational groups on management ideology.

The results presented in Model 2 also fail to support the idea that men are more likely than women to resist management ideology. Controlling for labor process, trusted worker, and unionization variables, men evidence significantly greater adherence to management ideology than do women. This finding may result from negative gender-based evaluations of (mostly male) managers by their female subordinates (see Cavendish 1982; Westwood 1984).

As expected, being a union member decreases adherence to management ideology. This is the largest and most significant coefficient in the model. Even after controlling for occupational group, labor process, trusted worker characteristics, and gender, unionization has a profound impact on worker consciousness.

## DISCUSSION

Technical workers are not as highly educated as professionals, but they are substantially better educated than other workers. Their occupational prestige is significantly less than that of professionals, but much higher than that of other workers. Technicians are not distinguishable from managers in either their educational levels or occupational prestige. Education and training differentials between technical workers and other occupations revealed in this analysis thus largely, but not completely, conform to prior expectations that technical workers form an intermediate group between workers and managers, on the one hand, and managers and professionals, on the other. Interestingly, the observed differences in the amount of time considered necessary by incumbents to learn their jobs suggest that, while technicians undergo substantial on-the-job training, they are not alone in this. Indeed, they report having approximately one year less on-the-job training than do either managers or professionals.

Technical workers participate less in organizational decisions than do managers. They do not believe that their promotional prospects are better than those of any other occupational group nor that their jobs are more secure. In terms of the nature of the labor process, we find that technical workers occupy, in some senses, an intermediate position between management, professionals, and workers. They have significantly less autonomy than do managers, but are not distinguishable in this regard from professionals and other workers. They consider their jobs more skilled than those of either managers or workers and roughly equivalent to those of professionals.

None of the trusted worker variables—participation, promotion opportunities, or job security—translate into greater commitment to management ideology, nor do any of the "labor process" variables increase adherence to management ideology. While there are occupational group differences in skill level and potential to use one's full abilities, these do not translate into adherence to management ideology. Technicians do not have significantly higher levels of autonomy than do other workers. It may be that although technicians perceive themselves as more skilled than other workers, their lack of autonomy and participation inhibits their identification with management.

Technical workers have very low levels of unionization. In this aspect, they resemble management, and we would expect this lack of unionization to lead to increased adherence to management ideology. Professionals in this sample are almost as heavily unionized as workers and we would expect this to lead to increased resistance to management ideology. Neither expectation is confirmed by this study. Professional workers subscribe to management ideology in spite of unionization while technical workers do not endorse management ideology in spite of their lack of unionization.

## CONCLUSIONS

Our findings provide little support for the thesis that technicians occupy the special position of trusted workers or that this status translates into greater acquiescence to management viewpoints. Technical workers, despite being completely nonunionized in this sample, do not subscribe to management ideology any more than do traditional members of the working class. Nontechnical professionals and managers exhibit similarly high levels of commitment to management ideology, while workers and technical workers consider management dispensable. Professionals thus appear to hold a much more elitist and management-oriented view of the division of labor than do technicians. Research on the political attitudes of the "new working class" reported by James Wright (1978) similarly concludes that technical and other workers share a general lack of confidence in the large corporations by which they are employed. It seems most accurate to consider technical workers as a distinct second category of positions (along with professionals), which has a certain degree of autonomy from the historic polarity between management and labor.

Our findings suggest that technical workers may have more in common with the working class on an ideological level than one would suspect from reading the later Gorz, the neo-Marxists, or the neo-Weberians. Both technical workers and nontechnical workers are resistant to management ideology. Whalley and Barley (Chapter 1) suggest that it is unlikely that technicians will be incorporated into the hierarchical structure of firms as corporate professionals. We agree with their suggestion that technical workers are more likely to approximate the position of blue-collar workers in hierarchies of occupational dominance. On this basis, technical workers, although distinct from the traditional working class in their level of training, appear to have an untapped potential for unionization. Technical workers in the survey analyzed here were completely nonunionized in spite of the fact that they are quite similar to other workers in their ideological distance from management. Unionization attempts have been relatively successful in recent years among professionals, particularly teachers, in spite of the relatively high adherence among professionals to management ideology. The untapped potential for unionization among technical workers may thus be considerable.

# PART II

# Studies of Technical Practice, Knowledge, and Culture

# 4

# CUTTING UP SKILLS: ESTIMATING DIFFICULTY AS AN ELEMENT OF SURGICAL AND OTHER ABILITIES

*Trevor Pinch, H. M. Collins, and Larry Carbone*

## INTRODUCTION

Skill is a component of technical work. In this chapter we hope to demystify one element of skill; we want to show how one component of skill works in practice and how this component can be transferred from person to person. The element of skill we discuss is learning how hard a given task is to perform, and how hard one should go at it before giving up. We will show how one can describe this knowledge in quasi-quantitative terms communicable to those who do not have the skill themselves. The assessment of the difficulty of any particular task turns out to be what one might call a "second order" quality of skill.

How are skills transmitted? The traditional or "algorithmical model" treats the transmission of skills as a matter of the transfer of bits of information, such as could be expressed in written form or even in computer programs. Within such a model, all elements of a given skill can be exactly specified. Written instructions, recipes, and detailed manuals are, in principle, all that are necessary for the learning of a new skill. The newer "enculturational model," on the other hand, holds that most of the time the transmission of skills is best seen as a process of socialization—like attaining fluency in a language (Collins 1974, 1993). Within this model, written instructions alone will never suffice in the learning of a skill. Skills must be learned *in situ* and certain tacit elements of a skill can only be passed on by direct instruction from a skilled practitioner.

The enculturational model, which was first developed in studies of science, has also had an influence on research into innovation in industry (von Hippel and Tyre 1993; von Hippel 1994) and artificial intelligence (Collins 1990). Recently, D. MacKenzie and G. Spinardi (forthcoming) have used the enculturational model of skill, which points to the importance of the continued existence of skilled practitioners as a resource in transmitting skills, to show that the ending of nuclear weapons testing programs means the demise of weapon making skills and perhaps the very fading away of the weapons themselves—at least the sophisticated versions.

The enculturational model may have been a success, but it cannot be the whole story. For instance, some elements of a skilled task can be spelled out explicitly. What counts as success in riding a bike, for instance, can be articulated algorithmically. For example, successful bicycle riding includes not putting your feet on the ground for support. Even the knowledge that a skill can be accomplished at all is useful in its learning. Yet again, such knowledge can be transmitted "algorithmically." In spite of the fact that each time a person learns a skill, a kind of rediscovery takes place, it is harder for pioneers because they do not know it can be done. To give an everyday example, imagine oneself encountering for the first time a spindly framework with two spidery wheels surmounted by a leather blade on a stick and not knowing that bicycles could be made to balance if ridden at speed—it would be hard to learn to ride. How much easier things are when one knows that a bike can be ridden! We call these explicit reflections on the possibility that a task can be accomplished "second order" knowledge of a skill. Thus, an element of second order knowledge about a skill is the awareness that the task can be accomplished. We discuss another second order element of skill in this paper; it is the question of how hard a skill is to learn and how hard it is to carry out. This information, too, can be transmitted algorithmically. To illustrate the point, we turn to veterinary surgery.

## FIELDWORK

We take veterinary surgery as our subject in many ways by coincidence; we could, we suspect, have observed any kind of technical work in order to illustrate our point. For a variety of reasons, veterinary surgery was convenient; it is also fascinating.

Our research on surgery has been conducted at a major university's department of veterinary medicine. One of the authors of this paper (Carbone) is a practicing veterinary surgeon and one of the operations discussed in this paper was conducted by him. Another author (Collins) has also observed surgical procedures on humans as part of a continuing project. The three authors have either experienced or observed some twenty-five hours of veterinary surgical procedures in connection with this project (of course, Carbone has engaged in many more hours of routine surgical practice than this). In connection with the project we have witnessed or participated in a variety of surgical procedures on both small and large animals; we have seen the same surgeon operating on different species and different surgeons operating on the same species, and the same procedure (neutering) being carried out on different species. Surgeons at different levels of training, including novices doing some of their first operations, have been watched and questioned. We have also observed other personnel, such as anesthetists and veterinary technicians. The methods used have included participant observation, unstructured interviewing and audio and video recording. We have also studied surgical training films.

## IN SEARCH OF THE UTERUS

Sometimes surgeons have to begin their work by deciding whether the anatomy of a patient is in accordance with the species' description in textbooks. Consider the following transcript of a surgeon attempting the "routine" spaying (neutering) of a female ferret. Note how what should be the familiar terrain of the ferret's body turns out to be a foreign land. (This, as it happens, was the very first operation we videotaped.) In the transcript, "S" is the surgeon, "SA" is the surgeon's assistant, and "R" is the researcher. Numbers in parentheses are pauses timed in seconds, e.g. (4) is a pause of four seconds' duration. Underlined dialogue is spoken with emphasis.

S: . . . so what I'm in search of is the uterus.
SA: I was just watching her breathing—she's looking good.
S: (noises) Okay *[picking up organs]* right here I have her urinary bladder, and these are some loops of intestines. *[changes instrument]* And assuming she hasn't already had her uterus removed, it should be just down below the bladder. (5)
Sometimes one of these little hooks will help me pick it up without having to paddle around too much in her abdomen. (4)
*[in ironic tone]* Sometimes. (5)
That's intestines. (12)
That's large intestine.
*[several grunts and poking around, changes instrument, the animal appears to urinate]*
(50) *[lots of searching, poking around]*
SA: Does this look familiar? We woke her up this morning—she didn't have time for a potty break before we got her over.
S: Yeah, yeah. (3)
What are you doing on me? (4)
This is not very promising. I'm trying to squeeze out some of her urine. (17)

So *[pointing with fingers]* that's colon and this is bladder. There should be a uterus interposed right between the two *[changes instrument]* and it's not . . . (22)
So this a kind of a search and destroy mission here, but I can't—

R: Yeah.

S: . . . find much to destroy at this point. So what I can do *[uses scissors]* is extend my incision a little bit further down and try to look for a couple of possibilities. One is that they're hard to tell at this age. That's bladder, more bladder.

R: What age is the ferret?

S: These are what, about four months old [SA]?

SA: Yeah.

S: What I'm trying to look for is evidence of a spermatic cord leading to the exterior that could be evidence of a male. (17)
Or the other thing I've been trying to see is whether or not below the bladder I have a little stump of uterus indicating that it has already been through surgery once. *[continues search]*

R: Yeah. (7)

R: It's not possible that it's a male ferret rather than a female ferret?

S: It is possible.

SA: They're pretty hard to sex at this age—at any age.

R: How do you sex a ferret?

S: That's colon.

SA: *[in overlap with S]*. It's possible by the anal-genital distance.

S: They are pretty much equivalent generally. Now I, um . . . (7). That's the ureter. So, um *[gesturing with hands as he explains]* the anatomy is fairly simple. There's a bladder and there's a colon, and in a true female, in between the two is the body of the uterus, and it comes up and splits into two uterine horns. *[resumes search]* But if there's no body to the uterus then there won't be any horns for us to see either. So *[tweezers in right hand, hook in left]* I'm afraid we may have another dud dud animal in terms of needing to be spayed. (5)
That's bladder, always a good landmark. (14)
I've only had this happen once before, one of our own technicians had a pet cat that she thought was a female. And we opened her up to take out the uterus and ovaries. And, um, there was nothing. And it turned out he was a male who had been neutered long enough ago that he wasn't very obviously male. And so . . . (3) *[puts instruments down]* Well, *[looking at SA—now 9 minutes 20 seconds after first incision]* I think I'm going to close it up and see if we can find testicles to remove instead, which come out through a different route . . . So . . .
*[discussion of stitching procedure]*

SA: So you're going to tell me when you want this male—

S: *[Giggly type laugh]*

SA: This *suspected* male, this *alleged* male ferret to [be anesthetized]
S: This next guy?
SA: Just let me know when you. . . .

In this case there is doubt right from the start as to whether the animal *can* be spayed. The ferret had previously been used in laboratory work and was to be rehabilitated as a pet. Its individual history was uncertain. Immediately preceding the operation, the surgeon had administered a sedative to a different ferret, only to discover after shaving the animal that it had been spayed already. Thus early on in the procedure the surgeon remarks that "assuming she hasn't already had her uterus removed, it should be just down below the bladder."

The surgeon then conducts an extensive search for the uterus, eventually widening his incision to look for more evidence, but we can see that the possibility that the ferret may not have a uterus to remove soon becomes salient. The surgeon remarks in an ironic tone (line 8) "Sometimes . . ." This hints that things are not going quite as expected. Again the surgeon reiterates his mapping, explicitly identifying the intestines and large intestines. After a lengthy period spent poking around inside the abdomen of the ferret the surgeon remarks, "This is not very promising."

For the third time he explicitly reiterates the names of the organs he has identified: "So that's colon and this is bladder. There should be a uterus interposed right between the two and it's not . . ."

Thus far, the surgeon has confirmed that the uterus is not present where he thinks it should be. He notes this in a way which nicely captures the surgeon's task: "So this is a kind of search and destroy mission here but I can't find much to destroy at this point."

As the operation continues we see the emphasis switching further from finding the uterus to trying to explain the anomaly. Has the surgeon indeed made a discovery that this ferret does not possess a uterus or has he failed to find the uterus because he has not exercised sufficient skill? More explorations are not capable on their own of resolving this question (for an analogy in science—"the experimenter's regress"—see Collins 1985). A surgeon can, of course, try to search for corroborating evidence. In this case the surgeon extends the incision further to look for indications of a missing uterus. He looks for the residual stump, or for signs that the ferret is in fact a male, or for evidence of an immature juvenile uterus which has not fully developed (a possibility not evident from the transcript but raised later in discussion with the surgeon). In this case he finds no such corroborating evidence, and concludes the search after one last attempt to find the missing organ.

Even at the point where the operation is terminated the surgeon cannot be certain that he is correct. Would a bit more time spent searching or the use of a different procedure or incision lead to success? Would a more competent surgeon find the uterus? Is he facing some rare anatomical variation?

Given what we have seen of surgery, the question arises as to how uncertainties are resolved in practice. Resolved they must be, for surgery must come to an end. Indeed, in surgery there is a compelling practical reason why such doubts must be dealt with quickly. The shorter the time the patient is exposed to anaesthetic and to invasion by germs, the better.

Uncertainty is resolved faster as surgeons grow in experience. This is why experienced surgeons are usually on hand when less experienced surgeons operate. But what is experience?

## HOW HARD IS SURGERY?

We asked a trainee surgeon who had just taken nearly three hours to complete his first spaying of a bitch (normally a routine task that should take around an hour) how he thought he might improve. He said that he had made certain mistakes of sequence, such as not tying the ligatures off soon enough and not giving the right priority to the problem of "bleeders" (small blood vessels which bleed into the wound)—next time he would prioritize better. He believed his manual dexterity would improve slightly on those elements of the job which required essentially routine movements. He hoped he would be more confident and relaxed next time, since he would know that he could do the job; this would enable him to recall the formal instruction he had received about this type of operation more clearly. He would also be less distracted by the novel aspects of the situation—the feel of skin and tissue, and so forth. He would, in other words, be able to apply general routines better and concentrate more on the critical aspects of the task at hand.

The above is, no doubt, only a partial list of the features of experience which merely begins to reveal how complicated a thing experience is. None of the elements of experience discussed in the last paragraph can be transferred from one person to another without the second person practicing the skill themselves. One will not experience a situation as less novel, nor be less nervous, nor be more dexterous in routine aspects of manipulation, simply by being told what one has to do. Such aspects of skill, however, are not the focus of this paper; we are concerned with what *can* be told.

For hardness to be a publicly available feature of skill, it must be expressed in a ubiquitous language. Though numbers are only understood in context (Kuhn 1961), the context is sufficiently ubiquitous for them to be thought of as invariant across many *local* contexts, or effectively decontextualized. The key is that hardness can be roughly expressed in terms of length of time or number of iterations that it takes to complete a task. For example, we tell our perplexed graduate students to expect to redraft an academic paper at least a dozen times—that's how hard it is to write a publishable paper (as opposed to a term paper, which may be redrafted but once.) That is one way of telling students what comprises the skill of writing a publishable paper, and resolving some of their uncertainty about what a publishable paper is.

Returning to the ferret's uterus, the generalizable features are the length of time and/or the number of iterations involved in the searching part of the anatomical "search and destroy" mission. While the surgeon has no previous experience of this particular ferret, or of looking for a missing uterus, he does have experience of finding the uteri of other ferrets. He knows from this experience that he should find the uterus almost immediately. In short, the surgeon brings to the operation some idea of how long it will take and how many times he should repeat the mapping procedure before giving up the search. We are not suggesting that these measures are exact, but measures do not have to be exact to be useful; the crucial thing is to know for sure when you are well outside the acceptable window of a search.

Needless to say, searches such as ferret uterus searches are initially calibrated in the surgeon's community. What is more, the bounds of the "window" will probably vary from community to community. For example, the size of the window might be different for human surgery and veterinary surgery. Once we have developed a window, however, we can say that if the surgeon spends over an hour looking for the uterus he knows he has spent too much time. If, on the other hand, he abandons the search after ten seconds, he knows he has not spent long enough. He also knows that one failed search for the uterus will not be enough, but fifty searches will be too many. From the video recording of the ferret surgery it seems that in this case between six and nine searches was considered adequate.

It is then on the basis of his understanding of the hardness of finding a uterus that the surgeon is able to tell when enough is enough. It is this, we suggest, which enables him to break off the search. This is something he could tell his inexperienced colleagues or set out in the form of written instructions.

## THE DISCOVERY OF TESTICLES

The learning of standards of hardness can be demonstrated by considering another operation which we witnessed. The operation was on a 15-month-old horse which was to be gelded—i.e., have its testicles removed. This was in fact only the second operation that we videotaped. We remark on this to emphasize that the kind of case discussed here is not rare. The horse was a "cryptorchid"—a condition where one or both the testicles have not dropped into the scrotum and remain inside the horse's body. In this case both testicles were retained inside. The condition is fairly standard and again the operation was expected to be fairly routine but, as we shall see, it was still less routine than removing a uterus from a ferret. The surgeon in this case was in the second year of his residency and near the end of his training (after four years of veterinary school and a surgical internship, his training would be complete after two years of residency). Because the surgeon was still undergoing training, a senior faculty surgeon was on call throughout the operation in case assistance was needed.

After the horse was made ready for surgery there was some excitement evident among the surgical team. The surgeon, on examining the horse, found the genital area to be unusually soft. Furthermore, around the prepuce there were observed to be a number of vestigial teats. This could be evidence that the horse had no testicles at all—that it was in fact a hermaphrodite. If this was the case the testicles would be combined with the ovaries in a very different location deep inside the abdomen.

An incision was made to begin the search for the right testicle. The surgeon (aided by an intern) commenced looking for the testicle by palpating the groin area—i.e. the area between the scrotum and the peritoneum (the wall of the abdomen). The usual practice in such operations is for the surgeon to search for the testicle by following the inguinal canal (the canal running from the abdomen along which the testicle passes in embryonic development) down into the abdomen. Because it is good surgical procedure to make the operation as noninvasive as possible, the surgeons start at the surface and work inward. Rather than looking for the testicle itself, the surgeons first look for a fibrous structure known as the gubernaculum to which the testicle should be attached.

All in all, the surgeon and intern spent thirty minutes searching for the gubernaculum in this region. The surgeons prodded with their fingers (blunt dissection), occasionally pausing to try and identify particular structures encountered. The skills exhibited were much more tactile than those used for the ferret operation. Often the surgeons could be seen to look away from the horse while feeling with their fingers deep within the incision.

At one point a structure was found which was taken to be the gubernaculum; however, there was no testicle attached and the surgeons concluded that they had been mistaken and that it was just some other unidentified fibrous structure. After thirty fruitless minutes the faculty surgeon was called in to help with the operation. Later the reason for this was explained to us by the surgeon:

> She has been more involved with more of the surgical procedures, with that one in particular and with all surgical procedures. She may have had a case similar to that before that she could rely on that experience to guide her in the way that she has done in previous surgery. I mean it's like, I guess, anything in life. When you do something and it works one time, you tend to follow that lead again until for some reason it takes you astray and you figure [it] out.

The faculty surgeon could not come immediately as she had to scrub up. Meanwhile the surgeon poked through the abdomen wall and immediately found the gubernaculum, by which he could pull out the testicle. The testicle was finally located thirty-five minutes after the first incision was made. As the surgeon told us, "And as soon as we poked through the peritoneum we could see the testicle . . . it must have been right there . . . We poked down right there. And usually if they're in the abdomen that's usually what happens." Ironically, the testicle was found just as the faculty surgeon entered the operating theater. There was much tension-relieving humor as the team joked that the faculty surgeon was so good that all she had to do was appear in the theater and the testicle was found.

The possibility that the horse could be a hermaphrodite was now no longer salient. This had been an idea in the back of the surgeon's mind, but was not a concern he was prepared to take seriously until all the more normal explanations for the missing testicle had been rejected. As he told us:

> Definitely it [the possibility the horse might be a hermaphrodite] was a concern. But I guess even with something unusual like this, we still tend to prove that it's not either normal, or within the normal range of findings that we normally have, before we try to put it into a category where it would be completely unusual.

After removing the right testicle the team then turned to the left testicle. Again an incision was made and the surgeon started to feel inside the opening with his fingers. In this case the testicle was found in a much shorter period—ten minutes—and with much less poking around and searching. The second testicle was found in an abdominal location, equivalent to that of the first. We asked the surgeon why the second testicle had been located so quickly compared to the first:

> Well, just like before, we did the right side first. When we approached the left side, when we got down into the canal, the testicle wasn't there and we could not identify a structure that was obviously the gubernaculum . . . So we elected to proceed straight down and poke a hole into the peritoneal cavity. And the other testicle was abdominal . . .

The surgeons seemed noticeably pleased at the time of the operation that they had found the testicle. They remarked to us:

> S: Did you guys get a good view of that?
> Rs: Yeah.
> S: It's beautiful. That was sitting right. It was like, right inside the abdomen. I just moved my finger around. *[To assistant]* (indistinguishable) If I don't see anything that looks like anything, I'm going to go right in. *[Laughter on part of surgeons]*

What the surgeon seems to be saying here is that he has learned something new. In future, when dealing with such cases of retained testicles, if he cannot find the gubernaculum immediately he is going to spend less time searching around in the groin area and instead will "go right in" (to the abdomen). This was a point which the surgeon confirmed in interview: "I mean in the future cases based on this one, if I couldn't identify the gubernaculum down in the canal, then I wouldn't waste a lot of time trying to find it. I'd go ahead and proceed right down into the abdomen to try and find the testicle."

So what has happened between the lengthy operation on the right testicle and the much quicker operation on the left testicle? The surgeon has learned that looking for a testicle in the groin area is actually much easier than he had previously thought. He now has new experience which enables him to revise his estimate of how hard a search for a testicle in the groin area should be. Rather than spending half an hour on such a search, he has learned that about five minutes is sufficient. Such a search is easier than he had estimated and in future he knows not to spend so long searching in that area before going on into the abdomen; the hardness, as measured by time, is low. No doubt this understanding will be tested in discussion with his colleagues. In telling one another "war stories" (Orr 1996) he and his colleagues will be communally learning and establishing a new standard of hardness for this part of the operation which could, no doubt, become a publicly available feature of the skill of horse-testicle location.

If we compare this operation to the one discussed earlier on the ferret we can see the similarities and differences. Both surgeons reached the conclusion that the object they were looking for was not present in the area in which they were searching. The experience base of the surgeons was, however, rather different. The surgeon searching for the ferret's uterus had spayed many ferrets before and knew just how easy that operation was to perform. The surgeon involved in the horse case had only operated on a small number of cryptorchid horses before and never on one in which both testicles were abdominal. In the horse case the surgeon did not know how hard the search would be until he had carried it out on the right testicle. By the time he moved to the left testicle, which he found relatively quickly, he had enough experience to revise his estimate of how hard looking for a testicle in the groin area should be. In future operations he would be guided by this new experience. His colleagues could both verify it and learn from it.

What we witnessed with these operations was the importance of estimates of difficulty of skills and how such estimates can be used and revised. We suggest that this generalizable property of skills—knowing how hard something is—is one of the elements of skill which enable surgeons to act practically in such a way as to resolve uncertainty in surgery. The hardness of a skill, measured in the ways we have discussed, is one kind of generalization which enables previous experience to be applied to new cases, both in the development of an individual's skills and the skills of a community; what is more, it is a feature of skill that can be made available to the as yet unskilled.

## CONCLUSION

We have suggested that the surgeon sometimes faces the question: "Is this organ really not here or am I too incompetent to find it?" This uncertainty is practically resolved, we have suggested, partly by the surgeon bringing his or her experience to bear. We have described a generalizable and publicly transmittable feature of experience which we call hardness but standards of hardness have, of course, a consensual element.

We have shown the way that a measure of hardness can help in the practice and learning of a skill. There is also the question of the difficulty of learning a skill; this, too, can be roughly measured. The hardness of learning a skill is not necessarily correlated with the hardness of practicing a skill. Thus, learning to play the piano may be very hard, but playing a tune, once you can do it, is easy (likewise learning to speak and speaking). On the other hand, learning to write academic papers is hard, and remains so (for most of us), even when we know how to do it.

We can now see that in our opening section, where we first introduced the idea of hardness, we did not distinguish between these two dimensions, while in the description of fieldwork we have talked only of the hardness of carrying out a procedure, not the hardness of learning to do it. Some skills are really much trickier than one would have imagined and it is useful to know that one will never master them without immense effort and practice (think of handsawing a piece of wood to an exact length). If one believed that one could learn to ride a bike in two minutes then one would conclude at the end of two minutes' effort that one was incompetent; thus it is important to know that learning to ride is hard, though it is worth noting that once one has learned, it is easy to ride. Again, if one believed that learning to play the piano could be accomplished in the same time as learning to ride a bike, one would soon give up one's piano-playing efforts. If, on the other hand, one believed it was as hard to learn to ride a bike as to learn to play the piano, the vast majority of today's bike-riders would never have taken it up in the first place.

We have tried to establish a series of points of decreasing generality. First, the enculturational model (or however it is referred to in other traditions), is hugely successful but leaves the use of information and the transmittability of more "formal" aspects of skill puzzling. We have shown how certain measures of skill can be used both in learning skills and in resolving uncertainties in their practice. This is not to say that these measures work outside of context, but some contexts are more ubiquitous than others. Since numbers are understood in very wide contexts, a measure of the hardness of a local skill can be effectively treated as a decontextualized "second order" feature of skill. We have shown how this works in the detail of two veterinary procedures. In doing this we have described the everyday working out of an aspect of skill; we have seen that the more skilled, among their other attributes, have more understanding of how hard it is to carry out a task.

We believe that many of the skills exercised in technical work are subject to the same kinds of consideration raised here. For example, technical workers engaged in repair tasks (e.g. plumbers) often have to price jobs by estimating the difficulty of the repair—itself a highly skilled task. Thus, learning the hardness of a skill is an important aspect of training the technical workforce.

# 5

# BLEEDING EDGE EPISTEMOLOGY: PRACTICAL PROBLEM SOLVING IN SOFTWARE SUPPORT HOT LINES

*Brian T. Pentland*

## INTRODUCTION

Hot lines are everywhere. Whether you buy a can of Coke or a computer system, manufacturers routinely provide a number to call in case you have questions, comments, or complaints. Although people seldom need help with their soft drinks, they do need help from time to time with their software and other complex technical products. As the complexity of the product increases, the nature of these calls and the level of skill required to respond to them increases. Soft drinks are easy, but software can be very hard. Even the best tested and most widely used software can break down in unexpected ways. For this reason, nearly every software product is accompanied by some kind of "technical support" to help users through the problems that routinely accompany the installation and use of any complex technical product.

The people who work in software support hot lines are members of a growing occupational community (Van Maanen and Barley 1984) of technical specialists who call themselves "support people." Support people do more than register complaints and disseminate product information; they are engaged in constructing solutions to customers' technical problems. Support people spend a great deal of their time interacting with customers. These interactions are often perceived by managers as an important part of the customer relationship and the financial success of the vendor.[1] Software support encompasses both technical work (re-

An earlier version of this paper was presented at the Academy of Management in Atlanta, Ga., August 9, 1993.

1. *The Economist*, March 2, 1991, p. 66.

pairing problems) and service work (repairing relationships). In this respect, support people are similar to the copier service technicians described by Orr (1996) and the radiological technicians described by Barley (1988b). While these occupational groups are superficially quite different, they share the common problem of troubleshooting complex machines in difficult and sometimes quite urgent situations. Indeed, the term "technical service" suggests this dual role for software support people.

In their role as service providers, support people occupy the boundary between the organization and the customer. In a sense, they mediate between the needs of the customer and the capabilities of the vendor organization (Pentland, in press). In their role as technicians, support people are engaged in creating and applying technical knowledge. Support people are constantly confronted with new situations and new problems, and forced to come up with new solutions. In this sense, support people occupy the boundary between the known and the unknown, between software that works and software that does not. When this boundary shifts unexpectedly, "leading edge" software can suddenly become "bleeding edge" software and support people are first ones called upon to fix it.

Software support is an interesting example of technical work precisely because it sits on the boundaries of the known, the unknown, the technical, and the social. Like many technicians, support people draw upon general, abstract principles to accomplish their work, but they must do so in situations filled with social and practical contingencies. What kinds of principles and procedures do support people use to solve problems? Is their knowledge base primarily "scientific," in the sense that it is drawn from abstract principles? Or is it primarily pragmatic, in the sense that it is drawn from experience? These questions are important because they have implications for the kinds of training and qualifications that members of this occupational group might ideally possess. They also have implications for our understanding of the nature of technical work in general. There is a tendency to stereotype technical workers as either applied scientists, who merely apply scientific knowledge, or as glorified craftspeople, who operate purely from tradition and personal experience. These stereotypes are difficult to sustain in the case of software support, however, because the work so clearly depends upon both.

## SCIENTIFIC KNOWLEDGE AND PRACTICAL KNOWLEDGE

Technical work is knowledge work; it is hard to imagine a competent practitioner of any technical occupation who is not, in some sense, knowledgeable. The term "knowledge" takes on many different senses, of course, and seems to come in many different forms (Machlup 1980). Burkart Holzner and John Marx (1979) identify several kinds of knowledge, two of which are especially relevant to technical workers: the pragmatic and the scientific. Pragmatic knowledge is based primarily on efficacy and personal experience: the test is whether something works. Practical experience is, of course, a major source of knowledge in technical work. Engineering knowledge, for example, has traditionally been based on pragmatic criteria, as have many medical procedures. Michael Mulkay (1984, 92) offers the example of a British surgeon using strips of pawpaw fruit to clear up a postoperative infection after a kidney transplant. The doctor could not explain *why* the tribal remedy worked, but he had seen it work before; his knowledge of this treatment was pragmatic.

Scientific knowledge, on the other hand, is based on logical arguments supported by objective evidence. Scientific criteria for truth have a strong grip on the minds of many scholars and academics. Standards of proof vary among fields, but the acceptable standard of rigor and reproducibility generally goes beyond a simple test of efficacy. One crucial difference between scientific knowledge and "mere" pragmatic knowledge is that scientific knowledge is explicitly intended to be objective or value-free. Scientific "truths" are believed to be independent of the particular interests or biases of the individuals involved in their production. Pragmatic knowledge, on the other hand, is subjective and value-laden. To say that something "works" implies that it works well enough for the purpose at hand, which may vary from time to time and observer to observer. Scientific truths, on the other hand, are believed to transcend time, space, and culture.

It is interesting to ask what kind (or kinds) of knowledge are important among software support people or any other community of technical workers. To the extent that knowledge is central to work, an understanding of what form it takes has many ramifications. First, it suggests what kinds of socialization and training will be most appropriate for the occupation. A community dominated by pragmatic knowledge would not be well served by scientific training, and vice versa. In the case of software support, it might imply that a computer science degree is (or is not) a necessary qualification for the job. Second, it suggests the kind of patterns we can expect in the growth of knowledge in a given area. Where scientific knowledge (for example, in medicine) grows slowly through double-blind, placebo-controlled clinical trials, pragmatic knowledge may grow more quickly through case histories and clinical experience, such as that of Mulkay's (1984) surgeon. Finally, the kind of knowledge in use has implications for the way in which knowledge can be stored and distributed within the community. To be of value, knowledge must be retained, distributed and applied within a social system (Holzner and Marx 1979). The ways in which this can be accomplished are constrained, to some extent, by the form of knowledge in question. By looking closely at the workers and their work, we can move past the stereotypes of the applied scientist and the craftsperson to get a picture of what Stephen Barley (1991) calls the "new crafts."

## METHODOLOGY

To gain some insight into the kinds of knowledge used in software support, I studied software support technicians in two different sites. One of the sites, Database International (DBI), is a vendor of database software for IBM mainframes. Their product, which I will call SystemTen, was used primarily by software engineers to develop large database applications. For example, SystemTen is popular among government agencies that need to keep track of large amounts of data. The other site, Advanced Publishers (AP), sells desktop publishing software for personal computers and workstations. Their product, which I will call FinePrint, was used primarily by writers and graphic designers to create books, manuals, newsletters, marketing materials, and other high-quality documents. These two sites were chosen to create as much contrast as possible in the complexity of the problems being solved. As a mainframe system development tool, SystemTen clearly anchored the high end of the scale. As a packaged application for personal computers, FinePrint was close to the low end of the scale. While the difficulty of the problems encountered at each site differed as expected, the analysis here will focus on the similarities between the two sites.

Data collection involved three months (five days a week) of observation and informal interviews at each site. Most of the time I sat with a support specialist, observing what he or she did and taking detailed notes. In some cases (particularly at Advanced Publishers), I was able to hear both sides of the interaction by using an additional headset or a speakerphone. For ethical and practical reasons, it was impossible to tape record calls. After each call was over, I would ask questions to clarify what had happened, what else might have been done, and so on. Unless the situation was urgent, people were generally happy to discuss their interactions. I would sometimes retreat to my office at each site to write up observations about situations where note-taking was not possible, and at the end of each day, I typed my field notes. Altogether, I observed 237 calls at Advanced Publishers and 159 calls at Database International, plus numerous meetings and informal discussions among support staff.

## THE BLEEDING EDGE

In the world of computers, one likes to be on the "leading edge," with all the connotation of prestige, power and speed. But when things go wrong, one can quickly end up on the "bleeding edge," a pun that I first heard from the support engineers at DBI. Even when systems are not completely state of the art, they are routinely subject to malfunctions and breakdowns, so much so that technical support and maintenance have become institutionalized within the industry (Pentland 1991a). Consider the maintenance process for IBM system software, as explained by one of the engineers at DBI:

Engineer: There are two schools of thought about maintenance [for IBM system software]. One school says, "Only apply exactly what you need, period." That way you minimize the risk of installing fixes with errors and introducing bugs into previously working code. Another school of thought says, "Apply all the maintenance, all the time." But even the people who say to apply all the maintenance all the time would *never* apply the most recent PUT level.[2] You always want to stay at least two or three levels back. You never want to be out there on the bleeding edge. If you're a few levels back, then you can order the bucket tape.

BTP: The bucket tape? What's that?

Engineer: The bucket tape has all the fixes for that PUT level. It's the fixes to the fixes that didn't work.

BTP *(incredulously)*: Is that a standard thing?

Engineer: Sure, you'd just call up and say, "Gimme the bucket for 8902." They've gotten a little tighter lately. They'll only send things that relate to your configuration, but they still have 'em.

A Program Update Tape (PUT) fixes old problems, but the "fixes" also create new problems. IBM has a "fix in error" rate of about 20 percent, according to engineers at DBI. Everyone knows that fixes create problems, so knowledgeable customers are wary of applying the most recent fixes to their systems. IBM issues "bucket tapes" for the courageous or foolhardy ones who venture too close to the edge and get hurt.

The difficulty in supporting complex, error-prone technology lies in locating the source of the error. One has to decide what is real and what isn't. In discussing a particularly tricky problem that took weeks to solve, a support engineer made the following comment:

> The possibility for bullshit is enormous in this kind of work. I think that part of the reason it took so long for anyone to solve that problem is that it was so easy for people to cook up theories, off the tops of their heads, and then spend days heading off down blind alleys. You can really waste a lot of time doing things that seem completely reasonable at the time. And all the more so because people's egos get involved.

2. A "PUT" is a "program update tape," and a "PUT level" refers to the most recent PUT that has been applied or installed. These are sometimes called "maintenance releases" and they are issued several times a year, depending on how many problems are found and fixed. "8902" would be the second PUT tape for 1989.

The point is that there are a large number of different ways that things can seem to go wrong, and a lot of time can be wasted figuring this out. In terms of the criteria proposed by Trevor Pinch, H. M. Collins, and Larry Carbone (Chapter 4), software support is potentially very difficult and complicated. There is a great deal of uncertainty concerning how much effort will be required to solve a particular problem, even when it appears, on first inspection, to be straightforward. Sherry Turkle (1984, 225) likens the debugging process to being stuck in a "colossal cave" with no idea where you are or what to do next. An engineer at DBI, complaining about customers who don't understand how complex some problems really are, offered this response:

> Some customers are so ignorant. They'll get a bad ZAP and then say something like, "It applied correctly, how could it possibly be wrong?" To which I'd reply, "How many grains of sand are there on the beach? Think about that for a minute, bonehead, because that's about how many ways it could be wrong."[3]

Sand on a beach is an apt metaphor for certain aspects of support work, such as reading through "core dumps" sent in by customers to document their problems. These "dumps" often include several megabytes of data printed on a twelve- to eighteen-inch-thick stack of paper completely covered in machine language, detailing memory locations formatted in hexadecimal notation.[4] A typical office might have several of these stacks sitting around on the desk, the floor, or the engineer's lap. To the savvy reader, these stacks of paper could provide valuable insights into the customer's file buffers, stack pointers, task control blocks, and so on. Watching the engineers at DBI sifting through the pages day after day, I could almost see the grains of sand slipping through their fingers.

Technical complexity is only part of the problem on the bleeding edge. Urgency compounds the difficulties, not only because customers often need a fast response, but because they are dependent on the outcome. Callers' organizations may rely on the correct operation of the software to conduct their business, as one engineer noted: "Our customers frequently have their entire businesses on

3. A "ZAP" is a set of instructions that replaces a section of code in a customer's copy of the software. It is written in hexadecimal, and may be very short (replacing only a few instructions) or quite long (replacing hundreds of instructions). A ZAP that replaces a single instruction looks like this:

```
NAME XXXXX MODULE
VER  23B8 4780 3070
REP  23B8 47F0 307C
```

Very short ZAPs like this are sometimes read over the phone, while longer ZAPs are sent via electronic media. In either case, engineers are required to execute the new code one instruction at a time before sending it out.

4. Dumps are also received on magnetic tape and microfiche.

our software. If their files are broken, it means their whole operation is in jeopardy."

Alternatively, the dependence may be personalized, as in this case in which a caller claimed he would lose his job if the problem were not fixed right away: "It's not just that my boss is gonna call your V.P. My *job* is on the line. If you can't fix this, I'll lose my job."

The metaphor of the bleeding edge captures this mixture of complexity and urgency, of needing to know but not being sure. As support people sift through the sand, trying to create an appropriate response to a given call, they are engaged in the form of knowledge work that is the essence of their job. To avoid getting completely lost and never reaching a solution, they must rely on some kind of knowledge to guide their search. The question is, what kinds of knowledge are most applicable in the conduct of this work?

## THEORY VERSUS PRACTICE IN SOFTWARE SUPPORT

As human artifacts, computer systems engender different assumptions than do natural systems. When we confront problems in nature, there is always the possibility that some critical aspect of reality may have gone unnoticed; we accept the "mystery of nature" as a given. Not so with computers, which are, in theory, perfectly deterministic and rational. The rational, deterministic quality of computers is what engenders, in part, the sense of "mastery" that Turkle (1984, 207–213) describes among computer "hackers." Our tendency to attribute these qualities to computers helps account for their power and cultural significance in organizational accounting and decision making (Boland 1987). Given this cultural background, one might expect software support people to espouse an idealized approach to problem solving. That is, one might expect the more experienced members of this community to see through the details cluttering a typical computer screen to the pure, logical beauty of the underlying machine. As we shall see, software support people generally believe that computers are inherently deterministic systems whose behavior can, in principle, be rationally explained.

If their work were purely technical and conducted in isolation from customers, support people might have the luxury of sticking to first principles. But in their role as service providers, software support people use contingent, pragmatic interpretations influenced not only by the technical features of problems, but also by the social features of the service situation. To see how this is so, we will examine two idealized principles that supposedly govern the operation of computers and software: determinism (programs run the same way every time) and rationality (any observed behavior can be explained in terms of the underlying program).

*Deterministic versus Contingent Behavior*

The attribution of determinism to computer systems is exemplified by the belief that programs do exactly the same thing every time you run them.[5] This belief was explicitly expressed by a systems engineer at DBI who told me, "Software isn't organic. It's not like it woke up on the wrong side of the bed this morning and decided not to run today. Something had to have changed. If you run the same program twice, it has to do the same thing twice." Determinism is axiomatic to support logic, perhaps even more so than in natural science. In laboratory experiments, results are expected to vary due to random measurement error, subtle differences in initial conditions, and so on. In digital computers, no such variation is expected.[6] To the extent that initial conditions can be specified, outcomes are perfectly determined.

The preceding quote also suggests a corollary to the axiom of determinism: if a program behaves differently than it used to, then something has changed. At DBI, "something must have changed" was almost a battle cry. One engineer described an idea for a support T-shirt, which displayed a prototypical customer interaction:

| *On the front of the T-shirt:* | *On the back:* |
| --- | --- |
| When did it start happening? | This morning. |
| What did you change? | Nothing. |
| What did you change? | Nothing. |
| What did you change? | Nothing. |

Calls where customers reported new, unexpected behavior while claiming that nothing had changed were not actually very common at DBI, but they were extremely salient because they violated the assumption of determinism. At AP, I observed an apparent instance of nondeterministic behavior when one of the support specialists was working on a problem. First the program worked, then it didn't, then it did, all without any plausible explanation. The effect on the group was electrifying. Everyone gathered around, offering hypotheses and explanations; the engineering department was consulted for advice; nobody could explain the "bizarre" behavior. When it turned out that the specialist was unwit-

5. The attribution of *behavior* to a computer system verges on anthropomorphism, but is meant only to imply "that which is observed." My use of the word reveals my near-member status with respect to my subjects, who routinely use this term and many others which imply intention and knowledge on the part of programs.

6. There are exceptions to this principle, but they help prove the rule. Telecommunications and data storage are aspects of computer technology where "error rates" are part of the ontology. But the roots of these "errors" are ultimately traced back to nature, to aspects of the physical world that engineers cannot perfectly control and must therefore make allowances for in their designs.

tingly running two different programs, one of which worked and one of which didn't, his comment was, "I sound like one of those pesky users, don't I?"

The assumption of determinism provides a background for reasoning about technical problems. For example, determinism figures prominently in the idea of reproducibility: if you can't reproduce it, then it's not really there. The logic of reproducibility was manifest in the protocol for escalating problems from the support group to the quality assurance group at Advanced Publishers. Unless a problem could be reproduced by the support group, the quality assurance group would refuse to devote any resources to it. The implication was that unless the offending behavior could be reproduced (and the conditions of its occurrence succinctly stated), there was no problem.

At DBI, the logic of reproducibility was also invoked, but there were many situations where it could not be achieved. In a sense, the ideal of determinism bumped up against the reality of the work. For example, many kinds of database operations are "data dependent," so the same program will produce different results at different times, as one engineer explained. "The same code run against the same file can produce two different results because it's data dependent. There may be one person whose birthday is the third this month, and 5000 [people with that birthday] next month."

As a result, the only way to reproduce certain behavior is to have the exact same data in the file that was there when the behavior was first noticed. This sometimes involves collecting "all the pieces" that caused the problem on the customer's site and bringing them to the vendor site, as in this problem at DBI:

> They say the raw journal causes the problem, but the problem they reported happened while using a merged journal.[7] I have no way of being sure of what the real story is unless I have all the pieces. If I'm going to recreate the problem, I need the exact same files they were using, and all the other [documentation] they collected when it happened. The problem I have now is that we don't have enough space on the machine to put all the files!

Since the data changes constantly in a high-volume transaction processing system and the original files may not be available, SystemTen problems can't always be reproduced in this way. When this happens, users have to create tests at their own sites, as one engineer described: "We're very removed from big production environments here. We don't work with large, multi-million-record files ourselves. So we have to depend on our customers to help us diagnose these kinds of data-dependent problems."

In addition to data dependence, support engineers at DBI deal with timing dependencies which are also difficult to reproduce:

7. A "journal" is a file that keeps track of changes to a database.

> We can't reproduce it [here] because it's a timing-dependent problem. It'll probably only show up in their application when it's being banged on by dozens of users simultaneously. The problem is a race condition which develops when the terminal goes into a transmit state, meaning that it's about to send data to the controller and doesn't want to receive anything. But if the controller already sent something, like a screenful of data, then you get a race condition . . .

Timing problems were fairly common at DBI because the systems are utilized by many users simultaneously and because the system is built for high performance. As one engineer put it, "SystemTen doesn't like to wait around. It's almost like a New Yorker, it's that bad." In several debugging sessions I observed, the exact millisecond at which events occurred was considered highly relevant and correspondingly difficult to reproduce.

The main exception to determinism at AP had a somewhat mystical tone to it. Occasionally customers would call complaining that they couldn't open a file or that the contents were garbled in some way. The answer most commonly given was that "your file must have gotten corrupted somehow."[8] The support specialists assumed that files occasionally get "corrupted" for unknown reasons. The process of corruption could rarely be reproduced, but the corrupted file was available for inspection and testing, and these tests *could* be reproduced. Thus, when a file became corrupted, it was clearly a problem because evidence of its corruptness could be reproduced easily (e.g., "I click on it and it won't open.")

Another class of problems at AP that defied reproduction had to do with errors made by callers who were careless or inattentive while using the sophisticated graphical interface to the product.[9] As one support person explained, the menus pop up so fast that the user often can't tell what happened. The difficulty in reproducing an "error" of this kind is precisely that nobody knows what happened. The caller knows that their document is a mess but nobody has any idea how it got that way.

While the ideal computer system is perfectly deterministic, the problems that support people confront are not ideal. There are innumerable contingencies that can undermine the pristine logic of determinism and reproducibility. In the next section, we will examine the theoretical idea that computer systems are rational and the ways in which this ideal is violated in practice.

8. This is an example of abductive reasoning which typifies software support: Symptom S is reported and since event E would cause S if it had happened, assume E happened until proven otherwise. In the case of corrupt files, it works like this: "Something wrong with your file? Hmmm . . . it must have gotten corrupted somehow."

9. The product used a direct manipulation or "point and click" interface similar to that found on Macintosh computers, except that the program "remembers" your last selection. It is a helpful feature, but the results can be very confusing.

### *Rational versus Pragmatic Explanation*

Rational explanation implies that the *explanandum* (the idea or entity to be explained) is connected to the *explanans* (the explanation) in a logically consistent manner. Pragmatic explanation eschews the need for any such connection and relies instead on the efficacy of the *explanandum*. In other words, the observed event isn't really explained at all; the fact that it works is enough. This distinction is made quite frequently in the literature on social epistemology (Goldman 1987; Holzner and Marx 1979).

At my field sites, it was widely believed that every aspect of a computer program's behavior could be explained by examining the code that produced it. In other words, a rational explanation explicitly linking the behavior to the code was, in principle, always possible. The connection between code and behavior is reflected in this engineer's remarks about testing:

> I have to argue that on scientific grounds, every part of the program is testable. It has to be, because somebody wrote it. I mean, it might be a little tiny piece of code that nobody ever uses, but as a matter of scientific principle, it can be tested. People wrote it, so people can test it.

Of course, if people wrote it, people also made mistakes writing it, and everyone knows that programs contain errors. The assumption of rationality doesn't ensure correctness; it merely states that the observed behavior of a program can be explained through reference to the underlying code.

The assumption of rationality is exemplified by several of my field observations. Consider the example of a customer who wanted to know the maximum number of records that could be put into a file. There were two authoritative sources available: the official SystemTen documentation, which stated plainly that the limit was 16.7 million, and the support person's boss, who also said the limit was 16.7 million. Her boss agreed that 16.7 million had always been the limit, but said that if Marsha[10] wanted "the definitive answer," she should talk to a developer. The next day Marsha found the "definitive answer" for why you can only have 16.7 million records in a file. It was because the file pointer[11] is 24 bits (3 bytes), so the largest one is FFFFFF (which in hex-talk reads "fox-fox-fox-fox-fox-fox").

10. Not her real name, but this engineer happened to be serving as "Marsha of the day," the backup to the customer service center, at the time this call was taken.

11. A "file pointer" is a storage location that keeps track of which record is currently being worked on. Each record gets a number, and SystemTen can only work on as many records as it can "point" to. Thus, the size of the file pointer limits the size of the file.

Marsha and her boss have over sixteen years' combined experience working with SystemTen, and their expert consensus was confirmed by the product documentation. Yet their highly authoritative answer was not "definitive." Only by finding the file pointer in the code and observing that it was 3 bytes long could a truly definitive answer be constructed. The limit on the number of records (an observable behavior) was explained by the size of the file pointer (a part of the code).

The belief in the rationality of system behavior was also manifest in various practices used to isolate problems or bugs. At DBI, there are "debugging tools" which allow a systems engineer to "step through the code" one instruction at a time to see how it "runs." Such tools are commonplace in software development and maintenance work. At DBI, engineers were required to execute new ZAPs this way before sending them out to customers. This was considered the ultimate test that the code was doing what you intended it to do.

Unfortunately, the ideal of rational explanation and testing has its limits. A product like SystemTen may be testable in principle, but too complex to test in practice, as one manager explained:

> I can tell you that SystemTen is off the scale in terms of complexity. There's no way on earth to create enough test cases for it. There's no doubt that it's beyond the realm of human understanding. Certain modules you just have to put a skull and crossbones on and treat as a black box.

At DBI, the product itself is too complicated to test completely. At AP, the product is more tractable, but the same problem arises from the multitude of different hardware platforms and configurations, plus the possibility of incompatible software. Customers frequently called with unique configurations of hardware, such as the customer who called about the new 21-inch color monitor he had for his Macintosh. Engineers in the quality assurance group at AP had tested the 19-inch monitor from the same manufacturer and knew of some problems, but had not isolated them to the monitor, the operating system, or the combination;[12] the support group had never even seen the new 21-inch version. Similar problems arose almost daily at AP. The standard solution was to tell customers to remove unfamiliar features from their computer until the AP product started to "work" again. No explanation of the offending behavior was needed as long as it could be suppressed. The urgency of the support situation contributes to this kind of pragmatism: anything that works is acceptable as long as it works *now*.

It can be difficult to construct a rational explanation for a problem even when

12. The results of this test were available to the support group as part of an equipment "certification" procedure whereby certain hardware and software combinations are tested for compatibility.

the symptoms are known. This was particularly true of certain errors that occur in complex technologies like SystemTen. Consider these comments from two engineers at DBI:

> Even if you have the dump, all you really have is a footprint in the sand. You can see that it happened, but you have no idea why it happened.

> When you get documentation on some of these problems, it's still not always easy to diagnose. It's like footprints in the sand. You know the pages are being allocated but you can't tell where.

In a metaphor that evokes the image of a detective investigating a dead body lying on the beach, the problem is all too obvious; only the reasons are hard to trace. In these situations, pragmatic solutions may be the only possibility. In one instance of an apparent timing problem, an engineer used a "diagnostic ZAP" to make the problem go away.[13]

> Their application runs perfectly with the ZAP we gave them, so they're happy enough. But to really figure out what's going on is going to be kind of ugly. The application runs perfectly with the ZAP, which is just as valid a piece of evidence about what the problem is as any other. It points to a timing problem with the subtask.

The actual problem was covered up rather than solved, but the customer was satisfied and the call was closed. As the quote suggests, getting the real answer would be "ugly," so a detailed explanation of what had actually caused the problem was never constructed.

The ideal of rational explanation is elusive for a variety of reasons. The detectives at DBI believe that behavior is traceable to the code, but *which* code? There were over 400 officially released patches available for customers to install, each of which revised some part of the code. In addition, there are hundreds of "custom ZAPs" which also modify the code being executed. Finally, there are so-called rogue ZAPs which are not officially sanctioned by DBI but which customers have sometimes installed. The problem with a rogue ZAP is that the support staff doesn't know it's there, and sifting through innumerable combinations can be a daunting problem. In one notable case, a rogue ZAP led to a problem that ultimately consumed several person-months of programming time to diagnose and repair.

13. A diagnostic ZAP is one that is intended to reveal the nature of a problem, not to fix it. These are used when there is a particularly serious problem that occurs only on the customer's system and no other means of diagnosis is available.

Support staff at AP conduct their work without being able to refer to the code at all. This is not a serious limitation because they are supporting a packaged application where the code never changes from customer to customer. Variation in behavior has to come from other sources, like the hardware, other software, the installation, and the document being created. But as we have noted, these elements may also be unavailable. Lacking the ability to construct a rational explanation, they are forced to fall back on "whatever works." After all, it isn't possible to construct a rational explanation if the *explanantia* are absent.

## DISCUSSION

In some respects, the logic of problem solving in software support is very similar to the logic applied to experimentation in the natural sciences: experiments are performed and results are observed. In theory, software support could be even more scientific than science itself because, as human artifacts, computer systems are perfectly knowable and controllable. The logic of rational explanation and deterministic operation can be applied freely and rigorously. This would lead one to expect that software support people are indeed cast in the mold of the applied scientist; but in practice, we discover a rather different set of considerations entering the picture, as shown in Figure 5.1.

In theory, behavior is deterministic and explanation is rational, but in practice, behavior is contingent and explanation is pragmatic. A program may behave the same way every time given identical initial conditions, but such conditions may not be attainable in practice. Likewise, even though a program is explainable in principle, such explanations may not be feasible. Thus the idealized technical rules are not always applicable to the particular instances that confront the software support staff. It is interesting to note that in spite of the differences in complexity of the products being supported, these generalizations applied to both sites. When software breaks down in practice, it stretches the limits of understanding of the people who must fix it, regardless of the complexity of the underlying system.

| | Behavior is: | Explanation is: |
|---|---|---|
| In theory | Deterministic | Rational |
| In practice | Contingent | Pragmatic |

FIGURE 5.1
Theory versus practice in software support

Given this analysis of the problem solving practices involved in software support, what can we say about the kinds of skills required for this work? As Pinch, Collins and Carbone (Chapter 4) point out, we know very little about how to characterize the kinds of abilities needed for technical work. Nonetheless, there are some tentative insights that one can draw from this study.

The most important insight is that local knowledge is crucial in software support. This is clearly the case in solving specific problems, because specific details are always potentially relevant. But the general principle flows from the contingent quality of the whole work environment. The most fundamental principles concerning computer operation (e.g., determinism) break down in practice. The idea that universal, scientific knowledge is adequate for competent software support simply does not make sense. At the same time, certain principles (like determinism and rationality) are important because they orient the problem solver and guide his or her expectations. As software support people grapple with the unknown, they venture out from the ideal into the contingent. Their theories of how machines ought to work provide a tether which keeps them from getting lost.

A closely related theme that emerged in each site was the significance of details to the problem solving process. There are details about the product (features, known bugs, etc.), details about the hardware (disk drives, monitors, networks, etc.), details about compatibility with other software (operating systems, telecommunications, etc.) and so on. It is worth noting that this sort of information would all be relevant even if one were simply attempting to solve one's *own* problems. When we start to bring the customer into the software-support picture, we discover that we need to know details about the customer's business, their hardware and software configuration, their particular application, and their priorities. Such information, if available, can enable a support person to arrive more quickly at an interpretation of the problem and a plan of action that meets the customer's needs.[14]

It would be misleading, however, to think of software support as simply keeping track of details. There is a great deal of systematic search involved in solving a difficult problem: constructing the scenario that led to the problem, identifying the cause, developing a solution or "work around" of some kind, and working with the customer to see that the proposed solution is workable and satisfactory. Much of this work involves the construction and sharing of narrative, as described by Julian Orr (1990) in the case of photocopier service technicians.

14. The need to keep track of these kinds of details gives rise to a large market for specialized databases that track customer problems. Each of the two sites that I studied used such a system, although during six months of observation, I did not see a single instance of a problem being solved simply by looking up the answer in the database. Old problems and their solutions occasionally contained clues, but there was always a considerable amount of sifting to be done before a new solution could be worked out. This is because the details of the customer's site (hardware, software, application, and business) were never identical.

The work, in that sense, involves building up coherent stories that relate closely to other, more familiar stories. By appropriately manipulating the plot, perhaps by proposing a certain "work around," the support person can bring the living narrative of an ongoing problem to a happy ending. Too many unexpected twists and turns, however, and the unhappy customer may demand his or her money back. The software support person is a kind of bricoleur (Lévi-Strauss 1968), who must tinker with the tools at hand to arrive at a result satisfactory to the customer.

The need to build and maintain customer relationships points to the critical need for social or interpersonal skills in software support work. This issue was raised in many different forms, explicitly and implicitly, throughout my fieldwork. For example, it was often mentioned that companies were unwilling to let software developers talk directly to customers (even though they had a greater depth of technical understanding and might be able to solve problems more quickly) because they lacked the necessary social skills. Support people must be excellent listeners, patient, empathetic, and above all, nonjudgmental. Users will make stupid mistakes, and support people should not add insult to injury. Furthermore, users are often angry when they call, so support people must be able to put some distance between themselves and their work and not take users' comments and reactions too personally.

The need for social skills flows, I think, from the pragmatic nature of service work. Pragmatic reasoning is inherently value-laden; the criteria for successful resolution depend on the needs and interests of the customer, as well as those of the support person (who is representing the vendor). The support person cannot simply declare a problem solved, because the objective is customer satisfaction. In fact, in both sites I studied, customer agreement was an explicit requirement for "closing" a problem report, thereby declaring it resolved. Since customer needs vary from time to time and from individual to individual, software support people must be able to discern those needs and, as best they can, meet them. Clearly, mastery of technical detail is not enough.

## CONCLUSION

The pragmatic character of software support arises from the confluence of technical complexity and social pressures to get the job done quickly and in accordance with customer needs. There is often not enough time or adequate resources to arrive at a deeper understanding of why something works; it is enough to know that it does. In software support, what is tried becomes true. Knowledge accumulates through the accretion of experiences in specific situations, but is always conditioned by the contingencies of customer needs and the changing details of the underlying technology. Theoretical principles help orient support people to the task at hand, but the ultimate test will always be, "Does it work?"

# 6

# COMPUTERS, CLIENTS, AND EXPERTISE: NEGOTIATING TECHNICAL IDENTITIES IN A NONTECHNICAL WORLD

*Stacia E. Zabusky*

## INTRODUCTION

In recent decades, the United States has witnessed an unprecedented rise in technical and professional occupations, a rise that promises to continue into the twenty-first century (Silvestri and Lucasiewicz 1991, Bishop and Carter 1991). The increase in the number of technicians and professionals suggests that we may be witnessing the "rebirth" of horizontally organized work in a political economy that has been dominated since the Industrial Revolution by a vertically oriented division of labor (Whalley and Barley, Chapter 1; Zabusky and Barley 1996). As with any birth, however, the resurgence of horizontal work forms is accompanied by significant pangs. In this case, those pangs are associated with the conflict that occurs when members of horizontally structured occupational communities (Van Maanen and Barley 1984) pursue their work in the vertical contexts of organizations. This conjunction does not represent a simple juxtaposition of two complementary forms; instead, it involves a contest for legitimacy, authority, and autonomy within contemporary organizations. The contest is played out particularly among those technicians who are coming to work in bureaucratic organizations in increasing numbers.

The contest is more than an existential one, since structures of work organization define our ideas, in a cultural sense, of what determines success and even

The work reported herein was supported under the Education Research and Development Center program agreement number R117Q00011-91, CFDA 84.117Q, as administered by the Office of Educational Research and Improvement, U.S. Department of Education. The findings and opinions expressed in this report do not reflect the position or policies of the Office of Educational Research and Improvement or the U.S. Department of Education. Special thanks go to Stephen R. Barley, Assaf Darr, Bonalyn Nelsen, Mario Scarselletta, and the attendees of the Conference on Technical Work held at Cornell University (March 1994), for comments, critique, and support.

of who is honorable or worthy (or, to link the cultural to the economic, of who is worth more). In general, people take the principles of vertically oriented structures for granted. In vertical contexts, legitimate authority is defined by hierarchical position. Those in superior positions are presumed to have greater knowledge than those in inferior positions and thus to have the capacity to direct the work of, and otherwise exercise power over, those in subordinate positions. The bureaucratic organization is the epitome of this kind of vertical division of labor (Weber 1968). Horizontally organized work, by contrast, is focused less on power and more on expertise. In a horizontal system, different groups of practitioners jointly contribute their distinctive efforts to the execution of work tasks (Whalley and Barley, Chapter 1). In horizontal work processes, collaboration, rather than command, is the key to getting work done. Horizontal divisions of labor appear in the form of communities of practice (Lave and Wenger 1991), such as those found among craftworkers, professionals (Hall 1986), and increasingly, technicians (Barley 1991).

The conflict between horizontal and vertical principles of the organization of work has long been a topic of sociological investigation. Indeed, sociologists have turned a great deal of attention to the problems that arise for organizations and practitioners alike when horizontally oriented work (such as professional work) is performed in vertically structured contexts (e.g., Hall 1968; Benson 1973; Harries-Jenkins 1970; Freidson 1986). This has been a particular theme of research on scientists and engineers working in organizations (Marcson 1960; Kornhauser 1962; Pelz and Andrews 1966; Ritti 1971; Derber 1982; Von Glinow 1988). Little has been written, however, about technicians as members of occupational communities executing their work in the contexts of bureaucratic organizations. In part, this is because technicians have been largely "invisible" in the sociological literature (Shapin 1989). Part of the reason for their absence lies in their anomalous status. Sociologists have had a hard time trying to identify the form of their work practice, in part because their work seems to transgress such firmly entrenched cultural boundaries as mental/manual, dirty/clean, and white/blue-collar work (Whalley and Barley, Chapter 1).

Despite this omission, recent studies indicate that technicians are experiencing the same conflicts in the workplace as other technical workers (such as engineers and scientists) have in the past. They, too, find themselves attempting to build careers and execute technical tasks in work settings which are often inimical to their own sense of how work should get done (Orr 1996; Zabusky and Barley 1996). Compounding this conflict for technicians is the fact that unlike scientists and engineers, they are typically accorded low status within their employing organizations. Thus, technicians are caught not only in an abstract, theoretical struggle between competing modes of organization, but also in a more concrete and immediate struggle for respect in their everyday work.

The ways in which these struggles play out in different organizations, however, are not uniform; even though a struggle for respect may be found everywhere, the issues differ. The variations lie in the ways the technicians' work articulates with that of the organizations in which they are embedded. As Figure 1.1 shows, the work of all technicians involves "manag[ing] an interface between a larger work process and the materials on which the process depends" (Barley and Bechky 1994, 88). Nonetheless, the way in which technicians carry out the "epistemic core" of their work in a social sense varies. Specifically, technicians can be either "buffers," substantively involved in the work process, or they can be "brokers," ancillary to that process but critical to maintaining the infrastructure which makes such work possible (Barley 1993).

Figure 6.1 represents the structural position of the buffer technician.[1] It shows that "the technicians' 'output' serves as 'input' for the work of an occupation classed as a profession" (Barley 1993). Barley and Bechky (1994) elaborated on the buffer model of technical work with an ethnographic case study of the prototypical buffer technician—the research laboratory science technician (other examples of buffers include emergency medical technicians and hospital medical technologists). The focus in this paper, by contrast, is on the brokers. Brokers are technicians whose work is "not substantively relevant to those who depend on their work"; instead, they are primarily "responsible for creating general conditions necessary for the work of others . . . [by overseeing] some aspect of the technical infrastructure on which a production system rests" (Barley 1993).

The critical yet ancillary role played by brokers is represented in Figure 6.2. This model shows that brokers' work cannot be defined by their employing organizations alone. The work of maintaining and developing a technological infrastructure requires broker technicians to integrate technological criteria and capabilities generated outside the organization with the needs and requirements of users of technological systems within the organization. To do their work, then, broker technicians must be simultaneously oriented in two directions toward the users of the infrastructure in the organization, and toward the developers of the technology in a wider and more diffuse technical community.

This dual orientation requires broker technicians to serve as linguistic interpreters of a sort: brokers are involved in repeated attempts to "translate" technological realities into terms users can understand, and, at the same time, to translate users' needs into terms that make sense in the technical world. Thus, even though broker technicians are formally employed by organizations, structurally speaking they appear instead to be positioned between the organization and the technical community, as Figure 6.2 shows. Technicians are thus "cutpoints" (Zabusky and Barley 1994, 5) in the work process; the position of "cut-

1. Figures 6.1 and 6.2 are reproduced from Barley (1993).

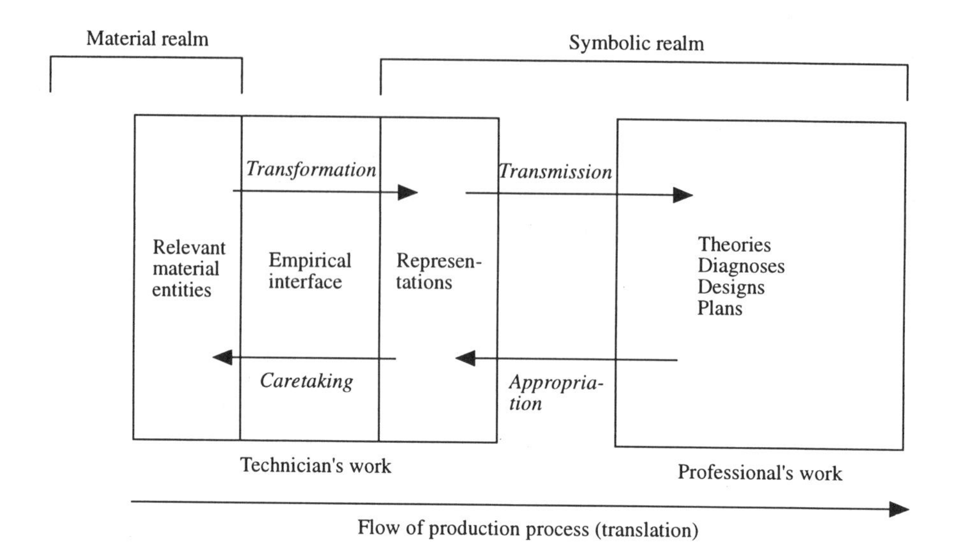


FIGURE 6.1
Technicians as buffers

point" suggests that technicians "serve as conduits for flows of information between . . . two communities." Were these "conduits" to be removed, there would be no communication between these two communities, with potentially adverse affects for the organizations involved.

The mediating position of "broker" complicates the technicians' roles as traditional employees in two ways. First, cutpoints potentially generate uncertainties of membership in a cultural sense—uncertainties which are manifest phenomenologically as problems of belonging. Do technicians and others in the organization view technicians as members of the technical community, the organizational community, both groups, or neither? What are the implications of membership for autonomy and control in the work process? The position of cutpoint also reveals that without these critical employees, organizations would have no way of getting the technological information they need to maintain their infrastructure. Their critical position is not matched, however, with high status in the organization. This incommensurability generates uncertainties of social status, notable particularly in the negotiations between technicians and organizational users about the proper stance from which technical service should be provided and received within the organization. These uncertainties of status manifest themselves phenomenologically as problems of identity: are technicians subordinates or partners in the work process?

Just such uncertainties define the day-to-day activities of microcomputer technicians working in large organizations today. In this paper, I focus on the cultur-

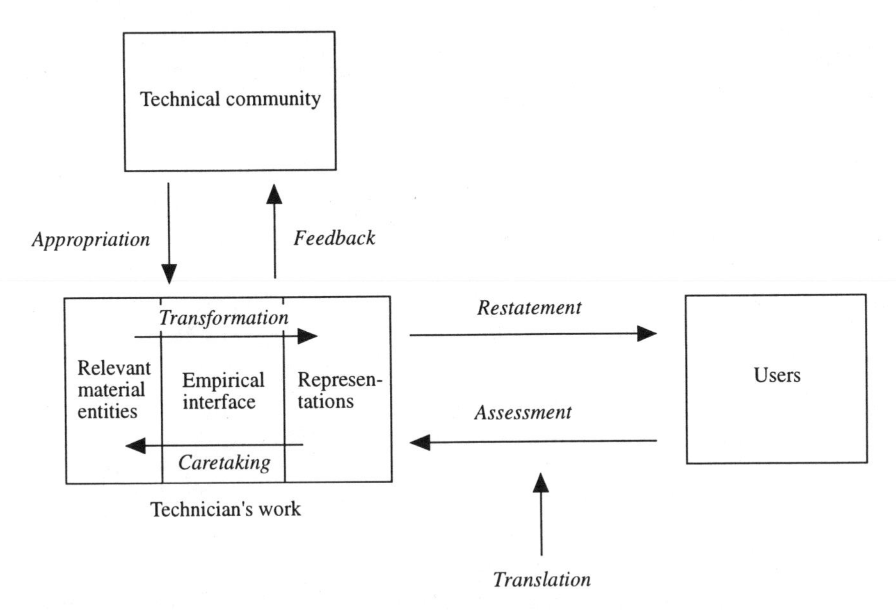


**FIGURE 6.2**
Technicians as brokers

al expressions of tension and ambiguity that accompany the mediating role played by microcomputer technicians. I show how these technicians and the users they support struggle to make practical and cultural sense out of the structural position of broker, a position which is neither explicitly acknowledged nor codified within the organization. These struggles are manifest in the way people talk about the problems of technicians' membership in the organization (are technicians insiders or outsiders?) and the problems of technicians' identity in organizational practice (are technicians servants or professionals?)

## RESEARCH SETTING

This paper is based on an ethnographic study of microcomputer support departments located in two schools (here referred to as the Research School, an interdisciplinary academic research school, and the Professional School) of a major research university. Data were collected over a period of about six months in the winter and spring of 1993.[2] Two to three days a week were spent on site; data collection methods included extensive observation and formal as well as in-

2. The data collection and initial analysis of field notes and interview transcripts were carried out collaboratively between the author, an anthropologist, and Stephen R. Barley, a sociologist, under the auspices of the Program on Technology and Work at Cornell University.

formal interviewing of technicians and users (the term "users" refers to all students and staff in the organization, including faculty, administrators, and clerical staff, who used the computer infrastructure).

Although the university where this fieldwork was conducted had a university computer department, responsible for directing and coordinating computer operations for most of the campus, and in charge of the electronic backbone which connected computers and phones campuswide, both the Research School and the Professional School had developed their own local area networks (LANs) which were independent of this system, and which were the exclusive responsibility of each School's computer staff. The local computer departments were, therefore, part of each School's organizational hierarchy. In each case, the head of the computer department (a computer technician or computer specialist) reported to an Associate Dean of the School. The organizational hierarchy, moreover, penetrated inside the computer department, at least formally speaking; in each case, there were internal supervisors (e.g., a technician responsible for supervising the work of other programmers or other workstation technicians).

Each School's LAN supported five to seven hundred users, for which each computer department employed a small number of technicians (eight in the Research School and seven in the Professional School). The computer staff in each department consisted of programmers, network administrators, and workstation technicians.[3] The programmers worked mainly with administrators (rather than with faculty or students). They were responsible for setting up administrative database systems, including developing report and query forms for clerical and administrative staff to use for putting data into and eliciting information from these databases. To develop database systems, programmers met with the users of the systems in order to determine the best structure for the database systems. For the most part, however, they worked alone, writing code and testing and debugging database systems. They were also available for answering questions and solving problems in relation to the daily use of these databases as they arose.

The network administrator was responsible for developing and maintaining the network system. This involved such routine activities as keeping the data servers from filling up, backing up the file servers on a regular basis, and "creating new users" on the system. It also involved more conceptual activities, such as determining how to organize the many users on multiple data servers, and determining whether new software packages would interfere with the operation of the network. Their work brought them into contact with the other computer department staff; they were also on the front line for any systemwide problems that might develop.

3. In addition, the Research School employed two professional statistical consultants, and the Professional School employed a researcher specializing in the internet.

Workstation technicians worked with everyone in the Schools, and engaged in a wide range of activities. They installed and maintained existing computer equipment (hard drives, monitors, keyboards, printers, etc.), as well as software on the network and in some cases on individual hard drives; they responded to users' complaints, questions, and problems, fixing software and hardware problems as they occurred; and they consulted with vendors and computer hot lines to troubleshoot particular problems. Workstation technicians were also involved in training new staff members on the use of the computer systems and some of the applications supported on the LAN. Their work was varied and unpredictable; they were more often out of their offices than in.[4]

## THE DAILY WORK OF COMPUTER TECHNICIANS

Although microcomputer technicians at the university did not describe themselves as brokers, when they discussed their work activities, they did articulate an understanding that their responsibilities demanded a dual focus. They talked about this duality not in terms of the role they played, however, but in terms of the activities in which they were engaged. On the one hand, they indicated, they were involved in "firefighting" and "housekeeping" work; on the other, they were responsible for "development" work. "Firefighting" and "housekeeping" described the part of their responsibilities focused on maintaining the computer infrastructure for the organization. This work entailed activities such as repairing equipment, upgrading software, keeping the network running efficiently, and monitoring and answering staff questions about hardware and software, diagnosing and solving problems as necessary. In a sense, the technicians were always on call. This meant that maintenance work was primarily reactive work, hence the term "firefighting." When engaged in such work, technicians were at the beck and call of users, since problems with computers at the workplace drove these activities.

In contrast, "development" work was more proactive. This aspect of the job involved researching changes in the market concerning hardware, software, and network technology; purchasing and testing new equipment and software applications in order to keep the infrastructure up to date; and researching, establish-

4. At the Research School, the many activities and responsibilities of the computer staff were complicated by the fact that the technological infrastructure combined Macintosh and IBM technologies. At the Professional School, the technology installed and supported by the computer staff was confined to IBM machines and software. This difference reflected, in part, the different historical trajectories of the LAN's development at each site, which in turn reflected the different organizational structures of the two Schools (more centralized at the Professional School) and the different user base of the two Schools (more homogeneous at the Professional School).

ing, and providing access to bulletin boards and databases on the internet. To do this proactive work, technical staff members read a wide variety of materials, including trade magazines and manuals, often on their own time. They also attended trade shows and academic courses, conferred with vendors' sales representatives and technicians at independent computer consulting firms, and read electronic bulletin boards on the internet to locate needed information and communicate with other people working on similar projects. In this aspect of their job, technicians acted independently of the users in the organization, since their own questions about technology drove these activities.

I would argue that technicians' views of the dual nature of their activities map onto the structural distinction depicted in the model of the broker role sketched in Figure 6.2 in the following way: maintenance work focused on the user/technician interface, and emphasized connections with the organization; development work focused on the technician/technical community interface, and emphasized connections with the occupational community. In these different activities, different concerns were correspondingly highlighted. In the reactive work which dominated organizational interactions, technicians were concerned with solving users' particular problems; here, the user's concerns, and by extension those of the organization, were first and foremost. In the proactive work which defined community interactions, technicians were concerned instead with discovering new technological possibilities. In this case, technical concerns were of primary importance.

To be sure, neither kind of work could be pursued in isolation. That is, the two worlds were distinct only analytically; to do both maintenance and development, technicians had to work at both interfaces simultaneously. Whether being reactive or proactive, whether dealing with a corrupted file (maintenance work) or testing a new graphics package (development work), computer technicians were always looking both ways. This was because technical changes had repercussions for organizational operations, and at the same time organizational requirements influenced the types of technical decisions which could be made.

Microcomputer technicians talked about this kind of simultaneous duality in various ways. At the Professional School, the computer staff said that "knowing the business" or "knowing the environment" was a critical element of their technical knowledge and skills. They insisted that organizational practices and needs be treated as an integral part of their technical work on a day-to-day basis. As one technician put it, they couldn't develop technological systems "in a vacuum" because the work of others in the organization affected their tasks. For this reason, at the Research School, the staff talked constantly about the importance of "communication" with users as critical to doing technical work well. Technical research undertaken in the absence of concrete organizational concerns would only result in inappropriate solutions. Thus, in order to solve technical problems, technicians felt that they had to be involved in "relationship work" in order to determine what information they needed to obtain in order to get the task accomplished.

When engaged in either type of activity, technicians were always negotiating this dual orientation in order to make practical, cost-effective, and technically feasible decisions about the operations of the computer infrastructure in their local organizational environments. This dual orientation—to the needs and interests of technical colleagues, on the one hand, and organizational users on the other—reflected their structural position as brokers. It also had a direct impact on their work, as they had to accommodate both aspects into the criteria they used for and information they incorporated into decision making and problem solving. More problematic, however, was the impact this duality of outlook had on their relationships with the users in the organization. In this case, the duality associated with being brokers in a structural sense led to organizational relationships characterized more by discomfort than by mutual regard. Indeed, the ambivalent assessments made by both users and technicians of the work and role of the computer staff in the organization centered on the situated meanings of doing maintenance and development work simultaneously, and the conflicting social commitments these implied.

## INSIDERS OR OUTSIDERS: ISSUES OF ORGANIZATIONAL MEMBERSHIP AND CONTROL

Given an understanding of technicians as cutpoints located in between two distinct work communities, it was no surprise to find issues of membership in the university focusing on where to draw social boundaries and what these boundaries might mean for control of the work process. The question of membership for the microcomputer technicians in the two Schools pivoted around the central issue of inclusion. Should technicians be included in the organization and regarded as integral members? Or did their outward focus toward another, nonorganizational social network obviate their membership in the organization? This was the sort of question faced by all social actors who found themselves in the in-between position of the cutpoint (for instance, the organizational scientists discussed in Zabusky and Barley 1994). At the Research School and the Professional School, the microcomputer technicians confronted and negotiated their rightful place in the organization on a daily basis.

Officially, the computer staff in both Schools were regarded unequivocally as members of the university (i.e., the local organization). In this sense, they were unproblematic "insiders." They were on the payroll, earned university benefits, had clearly delineated job descriptions, and were beholden to institutional evaluations and requirements when it came to raises and promotions. As described above, each School's management included the computer departments in its purview, and the departments routinely responded to administrative requests and followed management guidelines for budgetary spending and equipment purchasing. As members of the organization, moreover, the computer staff were assigned clearly defined positions and job titles according to the university's human resource guidelines. Some of the computer staff were formally designated as mid-level supervisors, responsible for overseeing the work of other technicians. The presence of these titles and formal lines of authority reflected the official view that computer technicians were unambiguous members of the university organization.

In interviews, administrators confirmed the official view that computer technicians were "insiders" when it came to accomplishing the larger organizational goals of the Schools in which they worked. Both Professional School and Research School senior administrators described the role of the computer staff as "critical" to the organization's mission, which included the production of research projects, papers, and books, and the education and support of students. The technicians were crucial in these endeavors because, as one associate dean at the Professional School explained, computers lay at the foundation of both academic and administrative work: "Computers here are tools we use to do our work; they are a means of getting work done faster [and] more efficiently." Moreover, as computers became an integral part of all students' and employees' "everyday work," everyone had come to depend on the computer infrastructure, expecting it to be available and functioning well whenever it was needed. The computer staff thus became more and more integral to the larger goals of the organization, since they were responsible for keeping the infrastructure going.

The associate dean of the Professional School further explained that the computer staff were critical to the students' and other staff's work not only because they performed necessary maintenance (which he referred to as their "support function"), but because they performed "an advisory function," in which "they provide[d] individuals with some idea of what they [could] use [the computers] for," in order to do their work better. To be able to guide users in this way, the dean concluded, the computer staff had to be, and were, "well integrated into the [organizational] system." From his managerial perspective, they were unequivocally "part of the staff" of the Professional School.

An associate dean at the Research School echoed this assessment of the importance of the computer in the research and educational mission of the organization. In his view, the computer had altered the work process in a fundamental way, making the computer technicians crucial to the work process. He described, for instance, how access to electronic mail had come to affect research and administrative work: "You get used to sending documents to people when working on a document jointly. You come in in the morning and you're [in a] hot mood, and you send out ten e-mail messages with four attachments to different folks." If the e-mail system was not working properly, the work process was disabled. Thus the dean acknowledged that the academic and professional staff had come to depend on the computer infrastructure to meet their institutional responsibilities. For this reason, too, the work of the computer staff could not be regarded as incidental to larger organizational goals, but had to be seen as a key component in realizing these goals. This dean, too, viewed members of the computer department as having "some sort of educational responsibility" to users as well as a more narrowly defined technical responsibility; in this way, he also viewed them as integral to the successful functioning of the organization.

Despite such official markers and managerial statements identifying microcomputer technicians as "insiders," however, ethnographic evidence pointed to the fact that technicians viewed themselves, and were viewed more informally by others, as problematic "outsiders." This evidence, collected from observations of day-to-day interactions and recorded in informal commentary, revealed the phenomenological difficulties associated with holding the structural position of cutpoint. Whatever the official view, computer technicians did not, culturally speaking, articulate any acceptance of their places as insiders. Instead, they talked consistently about the various ways in which they were outsiders.

Technicians certainly acknowledged the official markers of their organizational membership. They recognized that their paychecks had the university's name on them and they indicated that if they had a question about benefits, for instance, they would call the university personnel department. By and large, however, they considered this type of connection and affiliation irrelevant to the work they were doing. In response to the ethnographers' questions on this topic, for instance, computer staff members routinely downplayed their position as university employees. Many of them could not even readily identify who they worked for; when asked this question, instead of giving a direct and straightforward reply, they would pause, ask for clarification, and then dismiss the question with a wave of the hand or a laugh. The technicians dismissed the relevance of the university's supervisory titles in a similar fashion; most insisted that whatever titles they had were ignored by them and by their colleagues in the execution of their work. Finally, most of the technicians considered their jobs in the university to be interludes in long careers as computer technicians or programmers, careers which could and would take them elsewhere. They did not regard their careers, in other words, as linked to the organization, but instead as linked to an occupation.[5]

5. In Zabusky and Barley (1996), we argue that this occupational orientation in the pursuit of "careers" is not limited to microcomputer technicians, but is part of the experience and understanding of technicians in general.

The computer technicians emphasized their outsider role in describing their social interactions, or, more accurately, their lack thereof, with other staff members. They tended to experience, and even enjoy, a kind of social separateness. At the Professional School, the computer technicians made this clear right at the outset, when introducing the ethnographer to the physical layout of the Professional School. The computer staff showed her around, walking her through administrative offices, classrooms, and the library. In the process, they also talked about (but did not visit) the two staff lounges. When asked how often they went there, they laughed and said, "Never." The ethnographer asked with surprise, "So you're not really staff?" to which they replied with much irony, "No, we think we *are* the staff." Similarly, when asked where people might buy lunch at the Professional School, two of the programmers responded that they thought there was a cafeteria, but they didn't know where it was or who could use it, since in all the time that they'd worked there (more than a year for both), they had never visited it. In these mundane, practical ways, the Professional School computer staff revealed the extent to which they felt and reinforced a social separation between themselves and the other staff of the organization; in this worldview, the users were the insiders and the computer technicians were the outsiders.

The computer technicians also emphasized their outsider role in discussing how they were "different" from everyone else in the organization. This was a qualitative difference, something that identified them as inherently dissimilar to users. One computer staff member at the Professional School told the ethnographer quite explicitly that "computer people are different," because "they can see what the technology can do for people." They were, in a sense, visionaries, with an understanding of how transformative the computer technology could be for the organization and the users. On a more mundane level, another technician indicated that the computer people were different not only because they knew how the technology worked, but also because they were not afraid of it. This distinction, he felt, made it difficult for technicians to train users, because the technicians had to "think like somebody who doesn't know what they're doing"; somebody, in other words, unlike them. Some even made explicit their view that the ability to work with computers was an innate capacity; one said, "Some people can write songs; I write programs."

Technicians were not alone in their assessment of their outsider role. Analysis of organizational policies and interviews with users confirmed that, despite official affirmations to the contrary, most people did not view the technicians as bona fide members of the organization. For instance, organizational policy on space allocation enforced the social separation between computer technicians and the rest of the staff that the technicians themselves described. In both Schools, the computer departments were assigned to out-of-the-way places. The computer department at the Research School was located in a basement area of the main building of the School, a set of crowded, cluttered rooms in which dividers established smaller cubicles as work spaces for most of the technicians and programmers. The staff complained constantly about the inadequacy and inaccessibility of this space, seeing their placement there as indicative of the fact that their employing organization viewed them as outsiders whose needs were irrelevant to the organization's interests. At the Professional School, the computer department was divided in two; both offices were physically removed from the school's daily flow of work and life. The three programmers shared an office located in a small room off a ground-floor corridor traveled mainly by students going to and from classes. The door to this room, ostensibly for safety reasons, had no sign identifying it as the home of the computer department, and was kept closed and locked at all times, making the office virtually invisible. The other office, which housed the desks of the workstation technicians, the network administrator, and the internet specialist, was located in an even more out-of-the-way spot: four stories up, in "the tower," far removed from faculty and administrative offices as well as classroom space. One of the senior administrators indicated his belief that the placement of the computer staff in the tower and behind a locked door on the ground floor "where no one goes" was no accident; it represented the administration's effort to put the technicians in a place where they would not be in constant contact with other people. In this way, they were made into outsiders.

Users' talk about computer technicians revealed that they too regarded the technicians as outsiders. They articulated this view when they echoed the technicians' sentiments that the technicians were, in fact, decidedly "different" from everyone else. The users, however, did not put as positive a spin on this assessment as did the computer staff. Some even seemed affronted by the computer staff's inability to fit in. One administrator described the way "everyone who works in computers walks a different beat from everyone else at the Professional School." When the ethnographer asked what made them different, she replied, "They all don't take care of themselves; they don't shave, they have long hair—this isn't a criticism, just a comment." Some staff members revealed their distaste for this kind of difference when they called the computer technicians "computer geeks" or referred in a negative tone to the computer staff's overly "techie" focus.

The view of computer technicians as outsiders led to questions about the technicians' loyalty to the organization. Thus, this cultural perception had implications not only for organization membership, but for managerial relationships. The issue was framed in terms of trust and control: if the technicians were outsiders, could they be trusted to do their work in a way which kept the organization's interests first and foremost? And if they were, in fact, insiders, then shouldn't their work be subject to ordinary measures of organizational control?

From the technicians' perspective, the question of trust was unwarranted and the question of control was counterproductive. Their feeling of being different and their experience of social separation from others in the organization were not accompanied by a cavalier attitude about or indifference to organizational goals, even if they were accompanied by an indifference to organizational titles. For instance, the head of one of the computer departments told the ethnographer: "I'm *supposed* to report to [the Director of Finance], but he has no idea what I'm talking about." The technician added that he did not abuse this ignorance, insisting that "they put me in this position because they know me and they trust me." They relied on him to "come to work" as every employee should; in fact, he added, "if I didn't do anything for two weeks, it would be blatantly obvious . . . because things would fall apart." The administration relied on the technician especially, for instance, when it came to decisions regarding the appropriation of additional funds or other changes relating to computer infrastructure and technical staff. Administrators had to trust that the technician would use his knowledge of the market, the technology, and organizational constraints, to make reasonable and informed recommendations in these matters; in short, they were dependent on the technician's expertise.

A programmer at the Research School also talked about the fact that the administration had to have trust in the abilities and commitment of the computer staff, since the administration did not have the knowledge to assess the technical work of the technicians. She remarked that "we're perceived as specialists in areas where management doesn't have expertise." For this reason, she added, senior administrators did not say to them, "Now do this," but instead, "Find the best way to achieve this goal." In this way, managers set objectives but did not give orders. The technical staff's expertise, then, conferred on them considerable autonomy in the process of meeting organizational responsibilities.

This talk about trust on the part of computer technicians was, in part, an assertion of their dedication to the organization's mission; the technicians' focus on expertise, however, marked them as different. Computer technicians consistently talked about their sense of commitment in outsider terms. This was evident at the Research School, for instance, where computer staff members defined themselves as "consultants." As consultants, one technician emphasized, computer technicians viewed their relationship with organizational users as "the same thing as being an outside vendor coming in to repair something. You're repairing it for the client." In other words, computer staff members felt a responsibility to the work of the organization, but not to the organization *qua* organization. Being an insider was not, in their view, a prerequisite for being committed to the organization's mission. Similarly, assertions of loyalty to the organization were not necessary components of dedication to the work itself.

Administrators, however, worried about loyalty, wondered about trust, and often attempted to exert control in an effort to counteract such worries. This was in part because they were aware that the technicians did command a certain power, a power located in their apparent expertise with computers. One administrator at the Professional School explained that although managers "set the guidelines, they [the technicians] have control [since] they have the knowledge." The autonomy technicians experienced in their work, the administrator continued, was reinforced by the fact that, he had "no way to evaluate [their] work other than seeing the output and what I hear from other people." In short, managers had to trust that the computer staff would do their jobs, and do them well, on behalf of the organization.

The necessity of trusting the technicians, however, was at odds with the administrators' belief that technicians, as potential outsiders, could not be trusted. Administrators worried that technicians might put their technical interests ahead of organizational ones, an inappropriate prioritizing which would inhibit the ability of the organization to exploit computer technology in an appropriate and cost-effective manner. Such worries led to increased efforts to control the work of the technicians, despite the recognition by administrators that technicians' expertise made such control difficult. Such efforts to assert control were manifest, for instance, in administrators' talk about the "proper" role of technicians inside the organization. Administrators often insisted that it was position in the institutional hierarchy, combined with a clear commitment to the organization itself, which determined the right to make decisions about organizational issues such as hiring and promotion, budgets and resource allocation, and staffing priorities. In their view, only those in institutionally superior positions should be in charge of significant decisions, even in the domain of computers, since changes in the technical infrastructure had clear implications for work practices and departmental budgets. Although technicians might make technical recommendations, then, they were not empowered to make the ultimate decisions. When decisions were to be made, one administrator at the Research School told us he said to the computer staff, "We're [the administrators] going to do this. Now tell us what it costs. If we redecide [sic] after that, we will. But let's be clear who's going to make the decision." Thus, administrators asserted that the ability to exercise organizational control over the technological infrastructure lay in the institutional hierarchy alone.

These efforts to control technicians invoked sources of authority different from the authority recognized by the technicians. Administrators legitimated their power through the logic of hierarchical control, a form of authority consistent with participation in an organizational division of labor. Technicians, on the other hand, validated the legitimacy of authority only through expertise, a view consistent with participation in an occupational division of labor. It was not surprising, then, that assertions of authority by the administration engendered resistance and resentment on the part of technicians. They reacted to these attempts at control with renewed insistence that they were different because they had special expertise; moreover, this difference identified them as outsiders who should not be beholden to organizational lines of authority.

One way computer technicians articulated such resistance to managers' efforts to direct their work and thus transform them into unequivocal "insiders" was by asserting their power to exert what I call "technical revenge." When the politics of institutional control threatened to intrude on technicians' autonomy, a situation which happened repeatedly, the computer staff often responded with talk about how the technical "reality," which they identified with their own expertise, would undo administrators' efforts. For instance, on one occasion, the computer staff at the Research School was dismayed by the way administrators of the School had dismissed the results of a report they had drafted about potential changes to the LAN platform. The administration had decided to go against what the staff had proposed as the best option. The computer staff felt that this indifference to their efforts showed total disregard for their technical expertise, such that the decision was made entirely on "political" (that is, institutional) grounds. One technician, learning of the decision, joked to the rest of the group that the administrators responsible for the decision would suffer the consequences: "Let them see how slow it is; then they'll understand what we're talking about." In this way, he acknowledged the fact that technicians had a kind of power to affect decision making. This power was lodged in the facts that their expertise was not shared by others in the organization, and that they controlled the technological infrastructure on which the organization's work process depended.

Users in the organization also worried about the possibility of "technical revenge." To illustrate this, a computer staff member told the ethnographer about his predecessor, who had been fired. On the day the predecessor left, this technician had had to change all the passwords on all the systems in the Research School because the fired employee had had access to all of them as systems administrator, and the "user community" (that is, the staff in the organization) did not want to take a "security risk."[6]

The threat of "technical revenge" highlighted the way in which computer staff members' expertise was linked to their outsider status. University administrators, worried about disloyal outsiders within the organization, attempted to extend the reach of organizational authority in part by claiming that institutional position

6. It should be noted that the ethnographers never observed any actual instances of "technical revenge." In fact, we heard instead many expressions of an ethos of "responsibility"; technicians talked about how they would never use their expertise to undermine users' work. Many frowned on the activities of "hackers" who used their skills to break into computer systems or write computer viruses. One technician, after telling the ethnographer about how much he had hated his previous job as a programmer at another company, insisted that despite his dislike for his boss, he never would have considered sabotaging the programs he wrote in any way. This would have gone against his sense of professionalism.

was sufficient to dictate technical practice. The administration, however, did not have computer expertise, thus rendering their authority suspect in the eyes of the technicians. Assertions of control, then, had the effect not of bringing technicians further inside the organization's domain, but of reinforcing the technicians' sense that they were outsiders who did not belong.

As brokers, microcomputer technicians were cutpoints in the work process. Being a cutpoint might have made structural, analytic sense; it did not, however, make phenomenological sense within the boundaries of the organization. Members of the organization expected staff members to identify with the organization; if they were employees, in other words, they had to be insiders. But as members of an occupational community, technicians' presence inside the organization violated this one-to-one correspondence. Practically speaking, both technicians and other staff members recognized that their membership was not straightforward, that in a fundamental cultural sense they were, indeed, outsiders. It was this situation—of having formal employment but not cultural belonging—that led to the ambiguity about belonging that beset computer technicians in the workplace. Accordingly, they continually negotiated their membership and their status as insiders or outsiders, as trusted employees or untrustworthy strangers.

## SERVANTS OR PROFESSIONALS: ISSUES OF ORGANIZATIONAL STATUS AND IDENTITY

As suggested at the outset, questions of membership were not the only issues with which broker technicians were confronted as they did their work. Equally significant were questions about their status and attendant identities. Technicians and users argued, in a figurative sense, about whether technicians should be regarded as servants or professionals in the work process. This question was enshrined in a typical job title that some of the university computer technicians had had in other organizations: that of "support specialist." As "support" staff, technicians were identified as servants, individuals whose job it was to fix the organization's infrastructure as unobtrusively as possible. Moreover, as servants, they were subordinate members of the organization. As "specialists," by contrast, technicians were identified as expert practitioners, those whose job it was to solve problems encountered by others by drawing on their vast store of technical knowledge and skill. As specialists, they were on a par with other members of the organization, called in to help rather than to serve.

In this way, the title of support specialist placed technicians in an ambivalent space at the crossroads of vertical and horizontal divisions of labor. In the vertical division of labor characteristic of the organization, technicians were regarded as servants because their work was considered mundane and routine, subject to the control of those whose authority and knowledge were perceived as superior. In the horizontal division of labor characteristic of the occupation, technicians were regarded as specialists (or as "professionals," in the language used by technicians themselves and adopted in this paper) because their knowledge was integral to a work system which depended on the distribution and coordination of expertise. These possibilities intersected at the university, where the contradiction between defining computer people as servants or as professionals was marked in interpersonal relations filled with ambivalence.

Just how ambivalent this interpersonal terrain could be was made clear by a senior computer staff member at the Professional School who described the first time the computer staff had run "training" courses for all the School's users, a few years before. These courses were intended to introduce users to the network and to the use of various applications supported on the network. During the course of this training, the computer technician recounted, "it became clear what our status was . . . [as] interpersonally, people tried things they wouldn't try with someone they thought was professional." The problem was that "nobody could decide" whether the computer staff members were "librarians, janitors, or faculty." In fact, "it still comes as a surprise . . . [to many users] that people in this department can use words of more than one syllable."

This experience showed that users at the Professional and Research Schools tended to emphasize the "support" status of these staff members. In part, this was because when users and technicians interacted, the technicians were engaged in maintenance work. Users and technicians came into contact primarily when technicians came to troubleshoot problems, fix broken equipment, resolve software glitches, or get the network running after a crash. Users also expected technicians to be readily available to answer all manner of questions, no matter how trivial—questions which ran the gamut from how to do footnotes in WordPerfect, to how to send mail on the internet, to how to use the query command in a database program.

In interviews, users emphasized their view that the computer staff was there to "provide a service," specifically the service of "getting the tool working." This meant, as one user at the Professional School explained, that technicians should make sure the network, workstations, and printers were all "functioning properly." In providing this service, she went on, the most important element was responsiveness: "I would like to know when you send a problem to them, that they'd get back to you right away." She did not want to "sit around," waiting for them to deal with her problem. Users at the Research School echoed this same belief when they complained about what they perceived as the computer staff's lack of responsiveness. One user said, "I don't think they're too attentive to our problems . . . when your computer breaks down and you're in the middle of a rush job, you don't want to sit a day and a half."

For users, computer technicians' primary responsibility was to respond quickly and fix problems. Whatever ongoing research and testing work might be necessary to improve the individual technician's expertise, as well as the computer infrastructure, were totally irrelevant to users. Indeed, if anything, they perceived such activities as interfering with the computer staff's ability to be responsive. Because users valued responsiveness above all else, lack of responsiveness was often equated with lack of competence. One user at the Research School was blunt about his expectations of the computer staff in this regard: "What I want out of support is: make my machine work and leave me alone." From his perspective, if the computer staff could not do that quickly and unobtrusively enough (as it were, in the manner of servants, necessarily being seen, but preferably not heard), then that proved to him that they were not just lazy, but also "idiots."

Although users expected responsive service and complained when the computer staff did not provide it, they also recognized that the technicians had specialized technical knowledge which they brought to bear on the problems users encountered. Many were, correspondingly, grateful for and dependent on this expertise. Thus, users at the Research School could say that, although busy, the technicians were "very good" and "very helpful." But this dual expectation—that technicians be both responsive and expert—itself led to tension in interactions between users and technicians. In particular, users expressed their ambivalent expectations in the form of resentment of what could be regarded as technicians' displays of expertise. Such resentment seemed to suggest that users did not expect or want technicians to behave like professionals but preferred them instead to act like servants. Users were therefore irritated when technicians did not exhibit behavior appropriate to that status.

For instance, a user at the Professional School described in negative terms what happened when technicians came to her office to solve a problem (an event the ethnographers observed many times with other users). Before doing anything, she recounted, technicians first directed a series of questions at her, asking "Did you do this? Did you try that?" and generally quizzing her about how the problem had arisen and what she had tried to do herself to fix it. She resented this questioning, describing it as a kind of "interrogation"; she complained, "Sometimes they act like we should know all the intricacies of the computer [when they come to fix a problem]. If I knew how to fix it, I wouldn't be sitting here." She also took exception to the way the technicians subsequently "bored" her with long explanations of what had happened and what they were doing to resolve the problem, saying, "You just want them to fix [the machine] so we can do our work"; she had no need for "long explanations." Her colleagues, who were sitting by when she made these observations, concurred that they did not like it when the computer staff made them "feel stupid." The technicians' displays of knowledge seemed to contradict, for these users, their subordinate status.

Unlike the users, the computer technicians tended to emphasize the "specialist" status of their work in the organization. They preferred to think of themselves as professionals, and articulated this view clearly. They talked often about how they resented the fact that the maintenance work which formed a part of their duties came to consume so much of their time. Computer technicians bemoaned the overwhelming pressures of constantly having to "put out fires." Such work interfered with their efforts to pursue the proactive work of development, since they were constantly interrupted by phone calls or e-mail messages asking them to come and deal with what was immediately problematic. Since they were always on call in the organization, they found it difficult to pursue interactions with other professionals in their occupational community—a set of interactions which was, in any case, invisible to organizational users, even if it was an integral part of the technicians' work.

Computer technicians were in fact acutely aware of the low regard in which they were held by most of the organization's users. This made them uncomfortable with any type of "dirty" or "menial" work which users might interpret as subordinate or trivial. For instance, on one occasion, a workstation technician at the Research School was on his way to clean and fix a printer in a user's office. As he walked with the ethnographer to the office, carrying a vacuum cleaner, he remarked, "I hate this part of the job. I feel like a maid." The desire to avoid being identified as servants even led to efforts to hive off any such overtly "menial" tasks. At the Professional School, a work-study student had, in fact, been hired to clean and vacuum computer keyboards throughout the School.

That it was users' attitudes, rather than the work itself, which generated resentment on the part of technicians was made clear, however, by the technicians' otherwise ambivalent evaluations of such tasks. They were loath to declare the "housekeeping" work they did low or demeaning, since they did in fact have to do it. Instead, they tended to argue that even this "dirty work" was a critical and necessary part of the job. This meant that such "menial" tasks as vacuuming had to be performed with skill and dedication. For instance, at the Professional School, the ethnographer observed a technician fixing a dirty keyboard. He removed a key that had been sticking, blew carefully at the underlying mechanism to remove some dirt, and replaced the key, which now worked properly. He then surveyed the entire keyboard, which a student worker had recently cleaned. He was not pleased with what he saw, and remarked, "She didn't press down the keys well enough. . . . I know it's a boring job—I've done it—but even if it's a boring job, it has to be done right." With these words, the technician gave voice to his conviction that even the tasks of "servants" required skill, precision, and dedication.

In fact, computer technicians' primary strategy for evading ascriptions of servant status was less giving up their support role and the housekeeping work that went with it, and more emphasizing a different way of looking at the support they provided. It was not that computer staff members did not want to provide service; the issue was that they did not want to be regarded or treated as servants. Instead, they wanted to be respected for the unique talents and knowledge they brought to bear on the problems presented by the computer. They contested their identity as "servants" not by rejecting maintenance work altogether but by constantly evoking images of a horizontal division of labor, where computer technicians had higher status in the organization—a status commensurate with the critical work they performed. Rather than servants, then, they identified themselves as "professionals," in the minimal sense that they had specialized knowledge combined with commitment and dedication to do whatever it took to get the job done. Thus, when drawing attention to themselves as professionals, computer staff members referred to organizational users as "clientele" or "customers," people they were responsible for supporting (but to whom they were not subordinate) in a variety of ways.

When giving emphasis to their identities as "professionals," however, they insisted that they did not mean to ascribe to themselves any power based on credentials or institutional position. This, to them, would give unreasonable prominence to an organizational logic of hierarchical control and judgment, a logic which did not leave room for the respect of different talents which, in their view, were essential for getting work done in any organization. One programmer said to the ethnographer, "I don't know why some jobs are professional and some not. We're all performing functions which are important—would you like to eat on dirty plates?" A network administrator, in response to a direct question from the ethnographer, at first rejected the idea that he was a "professional," since at the Professional School, it seemed to him, "professional" identity was ascribed only to those with master's degrees or Ph.D.'s. In this view of things, he indicated wryly, he was only a "sub-professional" within a vertically oriented division of labor. From his perspective, however, he defined a professional more broadly, as "a person who does their job well, no matter what their job is." To demonstrate how broadly he intended his definition, the example he gave was of a garbage collector; if the individual did his job well, keeping his truck clean, not spilling refuse onto the street, and so on, then "he is a professional" because "it's not a matter of level or position."

It was no accident that these computer staff members used as their illustrative examples of professionals with expertise two types of occupations—food service and garbage collection—which are viewed by most people, and certainly by labor economists, as unequivocally "service" occupations: those occupations requiring low skill and having low wages, occupations in which people do for others what those others can do for themselves but do not want to. The computer technicians used these examples because they recognized the fact that organizational users often perceived them as servants. Rather than denying that they offered service, however, computer technicians resisted the implications of this identification by redefining its significance. In effect, they suggested that in a horizontal division of labor, no one could be considered a servant; instead, everyone was a kind of professional, in the sense that everyone had special skills and all skills were equally important. Indeed, some of the "dirtiest" tasks were in fact critically necessary to everyone's well-being—and for this reason alone, those who executed them were entirely deserving of respect.

This kind of discourse sought to flatten out illegitimate (from an occupational perspective) organizational hierarchies, emphasizing instead a division of labor in which expertise was distributed differentially, and in which mutual, collaborative work was the key to getting things done. Thus, one technician at the Research School described the support role they played, in not vertical but horizontal terms, as a "partnership with professors and staff." In this way, computer staff members articulated their conviction that organizational relations were to be governed by mutual respect rather than authoritarian control. A programmer at the Professional School, indicating that he did not like being ordered around by users, was explicit about the fact that he wanted users "to respect the work that I have to do and that I have other work to do. I can't just drop everything for three days [to attend to one user's problem]." The head of the computer department at the Professional School neatly summed up this difference between vertical and horizontal conceptions of relationships when talking about his promotion to his current position. Before he was promoted, he said, when faculty talked to him, they said, "Come in here and fix this computer"; after the promotion, it became, "Come in here and collaborate with me." This contrast was illustrative of the struggle of computer staff members in the Professional School and the Research School of the university to prove themselves as professionals in organizational environments where people repeatedly attempted to identify them as servants.

## CONCLUSION: PRACTICING A HORIZONTAL DIVISION OF LABOR

The ambiguities of membership and status which confronted the microcomputer technicians discussed in this paper were cultural expressions of the broker role these technicians played in the work process. These ambiguities manifested themselves in conflicts over questions of belonging, loyalty, and identity. In these conflicts, technicians consistently rejected the organization's definitions of what constituted work worthy of respect. They rejected users' focus on the values attendant on an organizational (vertical) division of labor by stressing their position as outsiders and their identity as professionals. What marked them as different was also what marked them as professional: their special focus on and expertise with computers.

In their emphasis on expertise and respect, computer technicians articulated a vision of organizational relations and work practices which challenged the prevailing models of organizing, supervising, and interacting in fundamental ways. Rather than an organization structured along vertical lines of authority, status, and knowledge, this vision emphasized a horizontal division of labor in which the organization would be characterized by myriad overlapping collegial and collaborative relationships. This horizontal structure was one which depended on the recognition, in a cultural sense, of everyone's expertise as equally valid.

This was no idle vision; the computer staff members were actively attempting to realize it internally in their own departments wherever and whenever possible. In these departments, the different skills and knowledge which members of the department brought to bear on the common and collective problem of the computer infrastructure was recognized as important. For instance, a programmer at the Professional School stated that "staff people don't realize the different functions [of computer people]; they think computer people can do everything. But I can't do everything [the workstation technician] does or [the network administrator] does; I come with a different perspective. I never wanted to get into the details of the hardware and the wires"; nonetheless, "all of these functions are important."

Similarly, a programmer at the Research School revealed her dismay at the way in which staff members "consider us to be all interchangeable . . . ; as far as they're concerned we're all down here doing computer things." In fact, she sometimes received telephone calls from staff members with questions not about database systems (her area of expertise) but about workstations and electronic mail and the like (questions she would, however, answer if she was able). She insisted, however, that "each of us has our own specialty. Part of [doing this work] is knowing when you don't know enough about something and you need to pass it on to someone else."[7]

This horizontal division of expertise, in the view of the computer staff, thus required the cooperation of all members of an organization. This cooperation depended for its effective deployment, moreover, not on institutional control but on trust, as everyone involved in such "partnerships" had to participate in a collective effort to solve problems and make the work system proceed smoothly. At the Professional School, the computer staff talked about their own internal working

7. The perceived interchangeability of computer staff members was reflected in the tendency of users to refer to these technicians generically as "computer people." In so doing, users emphasized again how different these staff members were from everyone else, as they identified them in terms of the machines that defined not only an outsider position in relation to the organization, but also a special domain of expertise.

relationships in this way. One programmer indicated that although he had an "official" position as supervisor of two other staff members, in practice this meant only that he filled out personnel forms every so often. His institutional position did not, in his view, reflect anything about superior ability or talents whatsoever, or give him the right to tell the others how to do their work.

Rather than managers who controlled the work process, then, this horizontal division of labor required coordinators who facilitated the work process. Such facilitation was necessary precisely because members of the organization had different sets of knowledge and expertise. At the Research School, one of the computer staff members with nominal supervisory capacity (that is, supervisor status delegated by the institution) spoke eloquently about the way this facilitating worked on the departmental level:

> It's actually very interesting looking at my position as opposed to Dave's [a pseudonym]. Because if you look at it from strictly what someone knows, if you know more you're higher up in the corporate ladder or whatever—that's definitely not true. I mean, when it comes right down to it, Dave knows a hell of a lot more about programming . . . I am willing to admit that he has more expertise than I do. And it would be foolish of me to try and tell him exactly what to do . . . I think if you're going to be an efficient manager in an area like ours, you've got to be able to acknowledge when someone else has more experience in certain areas than you do. And give them some credit for what they do know . . . I never have to say to Dave, "Oh, do this next. You're doing that wrong. Here's your work for today." That's not it at all. We sit down and we talk about what he's doing, what I'm doing, how they work together and we try to figure out the best way to tackle a job . . .

In the nontechnical contexts of the organization, however, computer staff members could not always communicate, let alone make real, their understanding of horizontal work practices. When they conceived of themselves as "consulting," for instance, others might perceive them as "serving" (or even, in rare instances, as "interfering"). In such contexts, they had to struggle to keep themselves from being submerged in those undifferentiated positions at the base of a vertical hierarchy of judgment and control. Moments of crisis, when a user called them urgently for assistance, provided them with critical opportunities to challenge these constraining organizational structures and roles. In such moments, technicians would experience a momentary inversion of the vertical hierarchy, as those relegated to the bottom suddenly found themselves on top. But these were only momentary lapses, and standard interaction and practice would resume as soon as the crisis was past. Nonetheless, such constantly recurring moments gave the technicians some measure of control through the execution of expertise in an organization which otherwise distrusted their role as outsiders and denied their identities as professionals.

By emphasizing a horizontal division of labor in which an ethic of collaboration prevailed, computer staff members articulated a vision of organizational relations distinct from that delineated by institutional rules and hierarchies. They made an effort to define the organization according to the practical problems and tasks which everyone was trying to solve and accomplish, rather than according to institutional status or "political" issues concerning personnel evaluations and the allocation of resources. They tried to work according to principles not of status hierarchy but of interoccupational teams, in which the efforts of collaboration were facilitated by a coordinator and not directed by a manager.

Organizations are feeling the pressures of these changes and of the presence of an anomalous set of employees ensconced at the very cutpoint of work systems. Managers, computer users, and computer technicians respond to these pressures with ambivalence and tension, tacitly recognizing in their practices what is not yet realizable in either organizational structures or policies. The brokers at the heart of these nontechnical organizations thus offer a constant challenge to traditional modes of organization through the ways they emphasize, rhetorically and practically, the belief that everybody—workstation technician, network administrator, programmer, secretary, faculty member, administrator, and janitor alike—is involved in a complex and elaborate collaboration that focuses on getting the work of the organization done quickly, efficiently, and successfully. In this, we see the shift to horizontal work forms before our eyes.

# 7

# WORK AS A MORAL ACT: HOW EMERGENCY MEDICAL TECHNICIANS UNDERSTAND THEIR WORK

*Bonalyn J. Nelsen*

## INTRODUCTION: VISIONS OF MORALITY IN SOCIAL ORGANIZATION

Generally speaking, two models—ideal types, really—have been used to divide and coordinate labor in Western society: the vertical and horizonal divisions of labor. Each model is marked by a distinctive moral vision which plays a subtle yet powerful role in shaping cultural images of work roles and relations. "Morals" are the rules or customs that regulate relations and prescribe behavior in a manner that enhances a social group's survival (Hall 1993). A moral "vision," then, is the complex of rules and norms which regulate interpersonal relations and prescribe modes of behavior in a given social setting. In vertically or "organizationally" (Whalley and Barley, Chapter 1) ordered social settings, roles are sharply differentiated and hierarchically ordered. Actors who hold higher positions in the hierarchy possess the authority to guide the actions of those below, an arrangement legitimized by the presumption that placement in elevated positions is indicative of superior technical and moral abilities.

That superiors in the hierarchy possess a greater measure of expertise is a fundamental belief underpinning the principle of authority. In fact, in hierarchical settings, superiors exercise authority legitimately only to the degree that their knowledge is believed to encompass that of their subordinates (Weber 1968). Authority is also underpinned by the presumption that superiors possess a height-

The work reported herein was supported under the Education Research and Development Center program, agreement number 117Q00011-91, CFDA 84.117Q as administered by the Office of Educational Research and Improvement, U.S. Department of Education. The findings and opinions expressed in this report do not reflect the position or policies of the Office of Educational Research and Improvement or the U.S. Department of Education. I wish to thank my colleagues who have provided helpful criticism along the way: Stephen Barley, Assaf Darr, Mario Scarselletta, and Stacia Zabusky.

ened sense of moral character, or "honor" (Weber 1968; Haber 1991). Simply put, moral character is the ability to foresee and appreciate problems as well as identify and assess alternatives, which entails both a capacity for objective, rational decision making and a willingness to assume responsibility for others. In vertical work systems, actors in the upper echelons are expected to act as moral compasses by establishing and enforcing a code of behavior conducive to the general interests of the collective. In return, actors inhabiting the lower echelons are expected to recognize and reward this effort by according respect, trust, and, most importantly, obedience to those above.

This assumption has laid the groundwork for the patterns of work roles and employment relations that have become uniquely identified with the modern bureaucracy. The ability to wield the power attached to any leadership role legitimately hinges directly on the presumption that actors holding such roles possess superior technical and moral qualifications. They are accordingly granted the right to command and, conversely, actors lacking these qualifications have imposed upon them the duty to obey. This notion has given birth to the patterns of superiority and subordination so characteristic of bureaucratic work systems, which happen to encourage social cohesion precisely because they have the effect of discouraging open, substantive debate about alternative means and criticism of ends.

Of course, hierarchically ordered institutions do indulge in other forms of leadership from time to time. For instance, innovations such as participative management, total quality management, and employee involvement initiatives entail a general relaxation of authority and less differentiation between the roles of superior and subordinate; but the fact that superiors ultimately retain the right to introduce, fund, coordinate, and revoke these practices at any time suggests that command and the power of sanction vested in command remain intact (Levitan 1984).

In contrast, roles within horizontally or "occupationally" (Whalley and Barley, Chapter 1) ordered work settings are neither sharply differentiated nor hierarchically structured. This arrangement is legitimated by the presumption that, upon attaining membership in a community of practice,[1] all actors possess roughly equivalent technical and moral abilities. A full member is, in principle at least, just as qualified to wield authority as any other full member. As a result, actors enjoy a measure of technical and moral equality which renders coordination by

1. A "community of practice" is a social group sharing similar skills, knowledge, values, language, and identity; in short, it is comprised of actors who inhabit a common social world (Barley 1993). Although commonly identified with groups such as members of occupations, communities of practice span occupational and organizational boundaries to include a variety of actors—provided they share the same worldview (Lave and Wenger 1991). For example, an emergency medical technician's community of practice includes firefighters and rescue workers, police officers, dispatchers, and a small handful of sympathetic physicians, nurses, and physician's assistants, in addition to the EMTs themselves.

command inappropriate. Horizontally ordered work systems therefore embody a moral vision wherein actors collaborate by means of persuasion and negotiation (May 1983). In such a setting, substantive debate about alternative means and criticism of ends is an accepted part of coordinating the relations of production.

The principle of command, however, is not entirely foreign to horizontal forms of social organization. Among crafts and professions—occupational forms which closely approximate this model—patterns of superiority and subordination exist for neophytes who have not yet become full-fledged members (Becker 1978; Groopman 1987). This is paradigmatic of the craft apprenticeship: apprentices willingly serve their masters only because they have not yet been initiated into the craft and hence are not colleagues. Once neophytes master the technical and moral standards set by the collective and thus achieve full membership, more collaborative patterns of association are the norm. Patterns of superiority and subordination are also employed when a member has failed to uphold technical and/or moral standards and must therefore be sanctioned by or expelled from the community of practice. This is witnessed on those rare occasions when a profession gathers itself to charge a member with malpractice. Although these occurrences are undeniably dramatic, they are the exception rather than the rule. Generally speaking, the moral vision characteristic of horizontal forms of social organization is decidedly collegial: coordination within or between communities of practice takes place through persuasion rather than command.

There is an obvious tension between these ideals; each reviles what the other demands. Yet, both moral visions are juxtaposed in the context of technical work. As members of hierarchical institutions and organizations, technical workers are subject to the strictures of command. Because they typically hold subordinate roles, a position presumably indicative of less technical and moral ability than that possessed by their superiors, employers and policy makers may freely impose rules and standards of conduct which seek to control the technicians' behavior. In turn, technicians are expected to acquiesce with a minimum of hesitation or complaint. This can be seen, for instance, in the experience of copier repair technicians. Julian Orr (1990; 1991a) found that employers sought to prescribe and control virtually every aspect of the technicians' work through the use of "directive documentation"—repair manuals, standard operating procedures, and so on. The premise was that following these prescriptions, which were crafted by engineers and blessed by management, would lead to the resolution of problems more quickly and reliably than could be accomplished by relying on the technicians' reasoning and know-how. When technicians deviated from prescribed practices because they believed the documentation was incomplete, their superiors viewed the problem as one of insubordination rather than technical inadequacy.

When viewed from the technician's standpoint, however, an apparent act of insubordination can take on quite a different meaning. Technical occupations approximate a form of social organization which functions by means of persuasion. The key to authority in this setting is not found in hierarchical "rank" or office, but rather in one's ability to make a convincing case for ideas and practices based on the contextual expertise and judgment that grow from personal experience. Hence, within the confines of the community of practice, it is acceptable for technicians to confer and debate with superiors, provided they have insight into the problem at hand. Indeed, empirical studies of technical work depict communities in which the normative relations of production are typically collaborative. Stephen Barley (1990) notes that sonographers regularly confer with radiologists during the diagnoses of ultrasound images, a practice warranted by the former's extensive experience with physiology and imaging technology. Similarly, Stephen Barley and Beth Bechky (1994) report that scientists and postgraduate assistants regularly seek the advice of skilled laboratory technicians when encountering problems with empirical procedures and physical phenomena. From this perspective, the copier repair technicians' deviation from prescribed practice is viewed not as an attempt to challenge managerial prerogative, but as a bid to negotiate an alternative means of accomplishing goals based on their contextual expertise and judgment.

It seems, then, that technicians experience a notable disjuncture between the moral imperatives that flow from their employing institutions and those that come from their technical communities of practice. To appreciate the implications of this disjuncture, one must note that the moral visions inherent in vertical and horizontal forms of social organization play an exceptionally important role in constructing our everyday understandings and expectations of work and those who perform it. They define social roles and shape the role incumbent's character by providing a compelling picture of the world and how one fits within it. They are powerful yet subtle forces which mold internalized standards of performance and obligation, ideas which are manifest in everyday speech and behavior.

Given this, we can see that failure to recognize and embrace the plurality of moral visions in the technicians' work world can spark at least two kinds of trouble. First, employers and policy makers who subscribe solely to views characteristic of vertical forms of social organization risk drafting rules and policies which can seriously impair the effective and efficient operation of the normative systems through which work is actually accomplished. If institutional directives are at odds with the rules fashioned by technicians themselves to guide everyday work practices, technicians must either negotiate an alternative course of action or be forced to choose literally between doing what is expected and doing what they believe the work demands. Because vertical and horizontal models of social organization project different and somewhat contentious images of technical work, the technician's social identity represents an uneasy amalgam of conflicting roles, norms, and obligations. These clashing images establish a contradiction between the technicians' self-conception and the ways they are treated (and called upon to treat others) which remains a constant and painful reminder of the anomalous standing of technical work in society.

In this paper, I will begin to explore the causes and consequences of this disjuncture of moral visions in technical work by examining the experience of emergency medical technicians (EMTs).[2] The case of EMTs is illustrative because not one, but two distinct sets of rules play a pivotal role in shaping the technicians' work experience. The first is mandated; it consists of the statutes, regulations, and standards imposed by various state, regional, and local authorities. The second is normative; it consists of the rules fashioned by EMTs themselves to guide everyday work practices. Although both share the same goal—giving patients the best care possible—each set of rules offers a very different definition of what constitutes "good care" as well as a different method of providing it.

This is no coincidence. Like that of most technicians, EMTs' work is embedded in institutional and organizational contexts which embody a vertical form of social organization. In contrast, their local communities of practice resemble horizontal forms of social organization. I argue that the EMTs' mandated and normative rules reflect moral visions characteristic of vertical and horizontal social settings respectively. Consequently, each vision prescribes a contrastive and somewhat contentious version of how EMTs should think, speak, and act in any given situation. EMTs thus face serious potential conflicts between the moral imperatives issued by regulators and supervisors and those derived from their community of practice.

## SITES AND DATA

The study was focused on two commercial and two volunteer agencies located in a single administrative region of a northeastern state. Each agency was subject to the same certification procedures, used nearly identical protocols, and answered to the same regulatory body. The first commercial agency, Jones, was a family-owned company located in a city of 50,000 people. Jones operated three ambulances and employed forty EMTs, including seven who worked full time. Medco, the second commercial agency, was run by experienced EMTs whom the company's founders had appointed to managerial positions. Situated in a city of 155,000, Medco employed approximately 100 EMTs to staff its fleet of nine am-

2. In doing so, I neither offer remedies for social ills nor draw on ethical theory to explicate the philosophical foundations of the problems found herein. Rather, I treat morals sociologically, that is, as objective social phenomena open to empirical investigation. In other words, I am concerned with "what is" rather than "what ought to be."

bulances. The two volunteer agencies were attached to fire companies in townships surrounding the city served by Jones. The first, Lakeview, operated two ambulances and was staffed by 18 active members. The second, Maryville, had twenty-eight active members and three ambulances.

The region in which the agencies were located recognized four levels of certification. Basic EMTs were trained to administer noninvasive first aid procedures such as cardiopulmonary resuscitation, oxygen therapy, extrication, bandaging, and splinting. Intermediate EMTs could establish intravenous lines, intubate via the trachea, defibrillate patients under limited conditions, and operate monitoring devices. By advancing to the level of critical care technician, an EMT could administer a variety of drugs and provide advanced life support in a wider variety of situations. Finally, EMTs certified as paramedics were allowed to administer narcotics and tranquilizers, apply external cardiac pacing, treat children under the age of three, perform emergency tracheotomies, and administer intramuscular and subcutaneous injections.

The data were collected in three overlapping phases during an eleven-month period of observation, during which I served as a participant observer in the four agencies. During the first four months of the study, I observed simultaneously at Jones and Lakeview. In the fifth month I began observing three formal emergency medical services (EMS) training programs that ran for a period of four months. Subsequent contacts with students and instructors facilitated entry into Medco and Maryville during the seventh month of the investigation.

I gathered data in each agency by attaching myself to a crew on duty at the time of observation. This strategy allowed me to experience the full range of activities that characterize EMS. Observations were spread over different times of the day and different days of the week to maximize the number of crews and types of calls observed. Periods of observation ranged from four to eight hours. Considerable time was spent in bunk rooms and parked ambulances, where informal interviews occurred spontaneously. Over the course of the study, I participated in a total of forty-three "runs" or "calls." All but two were with crews at Medco and Maryville, the two sites that officially permitted me to ride on their "rigs." Observations and informal conversations were supplemented by over twenty formal interviews with individuals and groups who interact with EMTs during their daily rounds of work activity; these parties included emergency room nurses, physicians, physician's assistants, regional EMS officials, radio dispatchers, and firefighters. Field notes were hastily recorded during observations and informal interviews and written up fully later in the day. Formal interviews were typically tape recorded. Where not otherwise stated, the excerpts in this paper are from my fieldnotes.

To highlight and contrast the effects of the moral visions characteristic of vertical and horizontal forms of social organization, I sought to identify the roles and rules experienced by EMTs both as members of an occupation in the health care hierarchy and as members of local communities of practice. I then collated all portions of field notes and transcripts in which EMTs referred directly or indirectly to rules or norms in discussions of proper or improper behavior.[3] These passages were read and iteratively coded to identify themes that defined the codes of behavior associated with being an allied health care worker and a member of a technical community of practice. Particular emphasis was given to themes emerging from the context of EMT training sessions and classes, settings in which rules were made relatively explicit for the purpose of socializing neophytes. These rules were then compared to identify any obvious discrepancies in the images and directives inherent in each. In addition, all instances of calls or stories about calls in which EMTs perceived a conflict between the rules and obligations inherent in each role were culled from the field notes. The notes were then coded for behaviors that marked the EMTs' responses to this conflict.

This analysis revealed that the technicians' work experience was indeed shaped by two distinctive sets of rules or "standards of care." The first was mandated; this consisted of the aggregate policies, standards, and rules imposed by the various state, regional, and local authorities charged with regulating emergency medical services. The second was normative; it represented the contextually specific code of behavior composed of what Robert Jackall (1988) describes as "moral rules-in-use," or everyday rules and norms used to guide social behavior in the workplace. The analysis also revealed the presence of some notable contradictions between the roles and relations prescribed by each standard of care. The remainder of this chapter will be devoted to a comparative discussion of these standards and the strategies used by EMTs to manage these contradictions.

## MANDATED AND NORMATIVE STANDARDS OF CARE IN EMS

### *The Mandated Standard of Care: EMTs as Physician Extenders*

Although paramedical personnel charged with providing emergency care and transport in the field have been a fixture of the military since the early nineteenth century (Barkley 1978; Howell et al. 1988), EMTs began to emerge as members

3. It is interesting to note that the topics of "ethics" and "morality" were not expressly discussed by the EMTs observed in this study, nor did I witness spontaneous discussion of the formal code of ethics endorsed by the National Association of Emergency Medical Technicians (American Academy of Orthopaedic Surgeons 1987). Indeed, relatively few EMTs knew of this code's existence, and virtually no one was familiar with its contents. However, the fact that EMTs were not lacking rules or standards to guide everyday behavior in the workplace calls into question the practical relevance of these concepts and the formal codes in which they are promulgated.

of an identifiable occupation only in the early 1970s. In 1966, the American public was transfixed by a government report entitled "Accidental Death and Disability: The Neglected Disease of Modern Society" (National Academy of Sciences/National Research Council). This report painted a sordid picture of emergency care and transport along the nation's sprawling interstate highway system by highlighting gross deficiencies in emergency first aid, pre-hospital and in-hospital emergency care, and trauma research (American College of Emergency Physicians 1984). Perhaps most shocking was the disclosure of the fact that a soldier wounded in Vietnam enjoyed a better chance of survival than a motorist injured along a U.S. highway.

Spurred by the public outcry which followed the release of this information, officials of the Departments of Transportation and Health, Education, and Welfare began to cast about for solutions to the problem of tending injured motorists. Although a handful of physicians had already pioneered the concept of public pre-hospital care and provided such care as a part of internships and residencies in selected locales, policy makers realized that, even if physicians could be enticed to practice in this nontypical setting, the development of a nationwide EMS system staffed by physicians would be wholly impractical for financial reasons. After a careful assessment of the types of care needed in most emergency situations, it was determined that a physician was not needed at the curbside; a mid-level practitioner trained to provide a limited range of treatment offered an alternative which was both effective and far more efficient.

As a consequence, the National Highway Safety Act of 1966 and the Emergency Medical Service System Act of 1973 had as their centerpiece a national system of emergency response which would utilize EMTs for pre-hospital care and transport. This legislation is notable for two reasons. First, it created an elaborate hierarchical system of federal, state, regional, and local authorities to plan, administer, and evaluate public EMS. From the beginning, Congress has been firm in the belief that sound EMS systems can be planned and operated and that funds can be wisely utilized only under the physician's watchful eye (Boyd 1983). Hence, the administrative bureaucracy responsible for designing, implementing and policing EMS has always operated under the general leadership of the medical profession. Second, the legislation established EMTs as physician extenders, or legal agents of the physician (Boyd 1983). As C. Eddie Palmer and Sheryl Gonsoulin (1990, 209) have noted, the EMTs' medical authority as physician extenders lies entirely within the emergency physician's license:

> Even though rarely at the initial scene of a medical emergency, the physician is vicariously and legally present in the eyes, ears, hands, voice, and knowledge of EMS providers. The physicians' power and abilities are thus extended and [their] abilities expanded. Again, the physician extender concept allows others to do the

work of the physician and the [medicolegal principle of] *respondeat superior* [let the master respond for the acts of subordinates] makes the physician liable for the actions of those "in the field."

Implicit in this principle is the notion that agents possess no skills or knowledge not also possessed by physicians, and lack the critical judgment and/or responsibility necessary for independent practice. Thus, the EMTs' practice must be circumscribed and closely supervised by physicians who remain responsible for their actions.

EMTs have offered an attractive solution to the problem of providing emergency care and transport. Over the occupation's 25-year history, the number of EMTs has grown steadily and their scope of practice has increased.[4] The EMTs' role in the emergency care process has gradually evolved from providing transport—essentially "feeding patients into the hospital system" (Sharp and Sharp 1989, 30)—to administering sophisticated medical assessment and treatment in non-hospital settings. Yet from its inception, the occupation has been dogged by controversy in both medical and regulatory circles (Palmer and Gonsoulin 1990). Much of this debate has turned on the issue of the EMTs' technical and moral competence.

Questions about the EMTs' technical skill and capacity for critical thinking are linked to training requirements. In an emergency context, EMTs do much the same work as a physician would perform, including fairly risky procedures such as cricothyrotomy, pneumothorax decompression, and cardioversion. Moreover, they do so in a context rife with chaotic, dangerous, and often uncontrollable elements with which hospital-based practitioners need not contend—and they accomplish these tasks with significantly less training. For example, physicians have at least ten years' worth of education and clinical experience (Schafft and Cawley 1987) while paramedics can spend as little as 600 hours in classroom and clinical training (American College of Emergency Physicians 1984). The EMTs' technical skills and critical thinking are expanded and honed by many hours of post-certification experience in actual emergency situations (Nelsen and Barley 1993); but because knowledge acquired in the classroom is more valued by society than the contextual knowledge gained in the field, suspicions about the quality of the EMTs' technical skill have been raised repeatedly by physicians and policy makers. In fact, some members of the medical community and EMS legislative bodies have repeatedly called for curtailing the EMTs' role in the patient care process (for example, see Smith and Bodai 1983, 1985; McCallion 1987; for a counter/argument, see Pepe and Bonnin 1989).

4. The number of EMTs and paramedics in the United States has been recently estimated at 500,000, although volunteers rather than paid EMS workers comprise the majority of this population (McQueen and Paturas 1990).

The EMTs' technical skills and decision making ability are not, however, the only sources of concern. That doubts about the EMTs' motives and responsibility continue to linger was made evident during my study when an operating room nurse who did not work closely with EMTs nevertheless insisted that "EMTs are loose cannons . . . you never know when they'll go off" and act in an impulsive, unpredictable, or irresponsible manner. These impressions are rooted in the historical origins of EMS as well as in EMTs' limited training requirements. For privately owned, commercial EMS agencies, images of profiteering stem from a past in which emergency transport was provided by funeral homes or limousine services more interested in turning a quick profit than in tending to the medical needs of their passengers (Peterson 1988; Palmer and Gonsoulin 1990).[5] Noncommercial status allows publicly sponsored EMS agencies to skirt this sordid imagery, but because public agencies are invariably linked to firefighting or police departments, the EMTs within them acquire a wild, rough-and-tumble image which portrays them as mavericks drawn to the work by elements of danger, excitement, or social control. Together, these images have contributed to the notion that EMTs' actions are motivated either by their paychecks or by the excitement of riding in a speeding ambulance, rather than by a compassionate concern for the patients' welfare.

The fact that EMS services have historically rested in the hands of volunteers in many areas also seemed to cast a pall of suspicion because some medical and administrative authorities considered volunteer EMTs to be less responsible than their paid counterparts. For example, a dispute erupted over allowing EMTs to administer narcotics in the area jointly served by Jones, Maryville, and Lakeview. Valium and morphine were approved for pre-hospital use by both state and regional agencies, and the EMTs were anxious to administer these drugs because they are very effective treatments for conditions frequently encountered in the field; but administrators in an area hospital were reluctant to authorize the drugs because they doubted that local EMS agencies, particularly volunteers, could be trusted to control their use. "The state can't make us carry a controlled substance," protested one administrator. "If the matter concerned organizations like [Jones], it might not be a problem . . . But in volunteer squads, simple things like controlling a drug box can be problematic because people don't show up for their shifts and a lot of people have keys to the drug box. And I won't be responsible when the stuff comes up missing."

5. This was the experience of Jones, a privately owned commercial EMS service which happened to be associated with a funeral home. During the field study, Jones's owner moved the service from the premises shared with the funeral business into a newly purchased and refurbished facility across town. Although it offered pragmatic benefits (more space, improved facilities, etc.), Jones employees suggested that the move was also undertaken to escape the unsavory and macabre imagery afforded by the agency's association with the funeral business.

In response to concerns about the EMTs' technical and moral competence, state and regional authorities have established a very exacting standard of care, a mandated code of behavior enacted to protect the public's rights and welfare. Of course, this is not unique to EMTs; state medical authorities establish standards of care for nearly all health care practitioners. There is, however, an important difference. When authorities mandate a standard of care for, say, physicians and nurses, the statutes, regulations, and codes thus created typically pertain to issues of licensing, accreditation, consent, negligence, and confidentiality (Schafft and Cawley 1987). The particulars of daily practice are left to the discretion of the practitioners. When authorities have drafted standards of care for EMTs, however, the particulars of practice have not been left to the EMTs' discretion. Rather, the standard of care is distilled into sets of protocols or written instructions which dictate practically every aspect of on-scene behavior. These begin with a patient's presenting condition and follow stepwise through the questions that should be asked, the tests that should be run, and the treatments that should be offered. Although broadly defined by each state, protocols are customized and administered by regional EMS agencies. They are locally enforced by supervising physicians and nurses, or "medical control," within the context of actual emergencies.

Figures 7.1 and 7.2 summarize the standard protocols for cardiac treatment used by EMS agencies in the region studied. In a very real sense, these protocols provide a virtual operative framework for structuring the EMTs' behavior by scripting the nature, order, pace, and frequency of communication and action during such emergencies. For example, the protocol in Figure 7.1 outlines procedures for routine care, or tasks to be performed in every suspected case of cardiac distress. It specifies what equipment must be brought to emergency scenes, the exact nature and order of medical procedures, the EMS certification level required to perform certain tasks, and when medical control should be contacted. Figure 7.2 depicts the protocol which is used to treat a specific cardiac ailment: ventricular fibrillation and pulseless ventricular tachycardia. These further specify which procedures may be performed without a physician's permission, who can be treated, what information EMTs must give to physicians, who may give medical orders, and when to transport the patient. Not listed but still enforced were rules regarding where a patient was to be taken for treatment, the equipment required on an ambulance, the documentation to be completed for each patient, and the overall amount of time allotted for all medical procedures and transport.

Protocols seem to provide an elegant solution to any problems regarding the EMTs' technical and moral competence. Because EMTs apply them under the direct supervision of physicians, any shortcomings in skill or knowledge can be quickly detected and corrected. Moreover, because the methodical, regimented

FIGURE 7.1
Protocol for routine medical care

**The following procedures will be performed on all medical emergencies requiring Advanced Life Support:**

- Advanced Life Support equipment will be brought to the patient. This will include a monitor/defibrillator, advanced airway equipment and oxygen, suction, medication box and capability for paramedic-to-physician communications.
- Reassurance and proper positioning of patient.
- Patient assessment with vital signs every five minutes and/or after every treatment.
- Airway management, ventilatory assistance as needed, and oxygen therapy.
- Start I.V. and D50W[a] running KVO[b] unless otherwise specified.
- EKG monitoring.
- Contact Medical Control within twenty minutes.

**Intermediate Life Support procedures**

- ILS providers will start IVs on medical emergencies only under the following conditions:
  —— The ILS agency is a non-transporting rescue unit awaiting arrival of a dispatched Advanced Life Support (ALS) unit.
  —— The ILS unit will be transporting and the IV can be started en route.
  —— Per Medical Control order.
  —— Under direction of ALS personnel.
- ILS units will meet an ALS unit or transport to a hospital, whichever brings ALS to the patient faster

**Multiple patient procedures**

- If two or more ambulances or three or more patients are involved, this represents a potential multi-casuality incident and the Resource Hospital should be notified as soon as possible.

[a]"$D_{50}W$" is the abbreviation for a solution of dextrose, measured 500 grams in 1,000 milliliters water.
[b]"KVO" is the abbreviation for the directive "keep vein open," indicating a slow drip rate.

approach to care encoded in protocols obviates much of the need for on-scene decision making about patient care, the chances that an EMT will commit errors in judgment are greatly reduced. Protocols also clearly establish the EMTs' social responsibility as health care practitioners. Simply put, EMTs are legally obligated to follow protocols. If they believe the administration of a specified treatment to be ill advised, they can contact medical control to request alternative orders. However, if they are refused, EMTs have a duty to comply by administering the mandated treatments. Because they are physician extenders who hold no medical authority independent of the physician's medical license, failure to follow protocols constitutes practicing medicine without a license—a felony offense. Violation of these rules may result in a punitive review of the action before local, regional, and, in the case of flagrant disregard, state authorities. Penalties may include formal reprimands, temporary or permanent loss of certification, fines, and imprisonment.

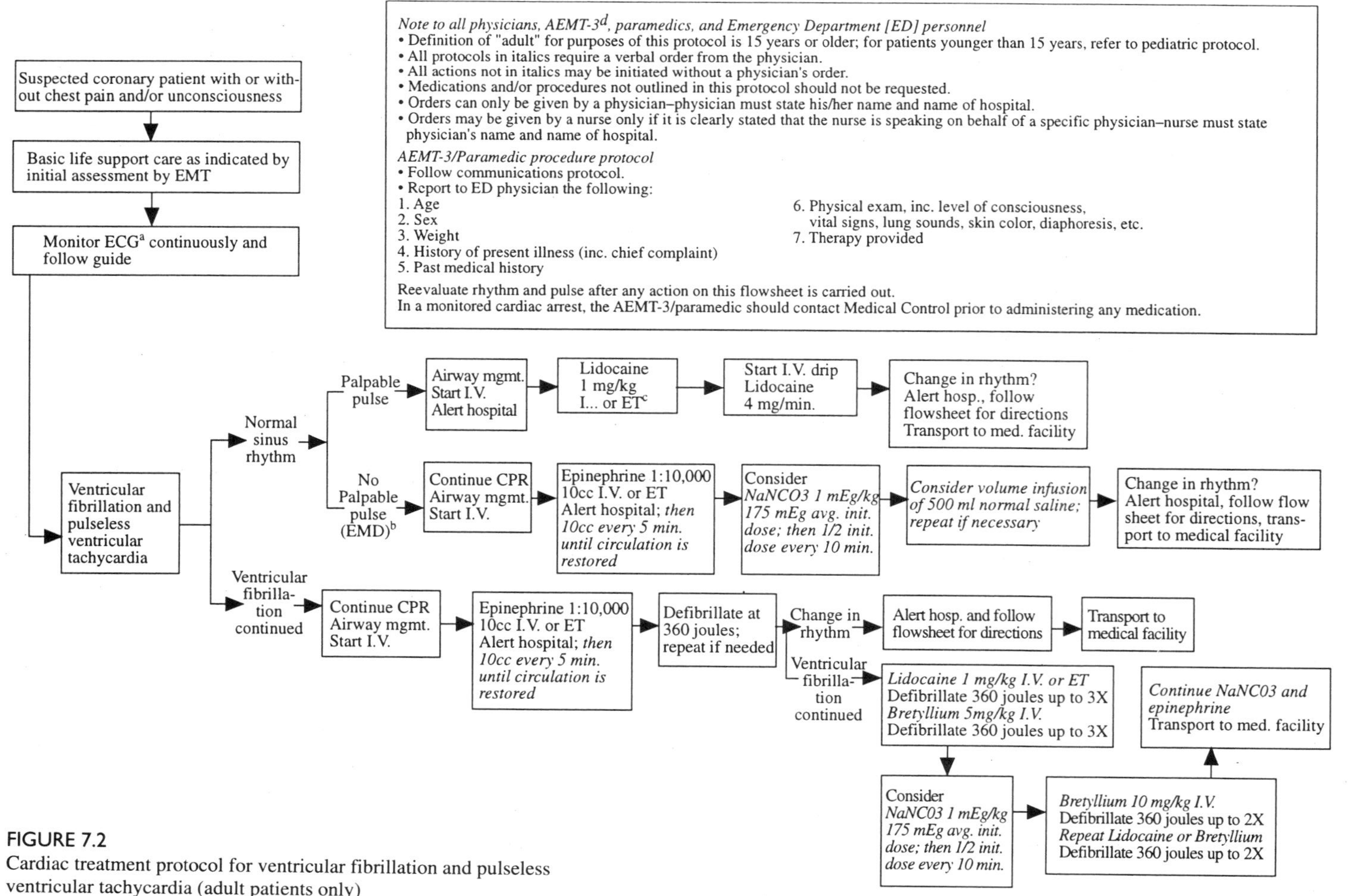


FIGURE 7.2
Cardiac treatment protocol for ventricular fibrillation and pulseless ventricular tachycardia (adult patients only)

### *EMTs as Partners in Emergency Care: The Normative Standard of Care*

The EMTs' mandated role as physician extenders stood in sharp contrast to the normative role assumed in the context of everyday work. Rather than serving as mere extensions of the physician, EMTs considered themselves to be partners in the emergency care process and were adamant about experiencing a more active role than their lowly status as physician extenders would seem to allow. It would be tempting to interpret the EMTs' insistence on parity with other health care providers as a bid for political power. Sociologists of work and occupations have long depicted the health care community as a hierarchy thoroughly dominated by the profession of medicine; therefore any gains in skill, power, or status acquired by allied occupations must presumably be made at the medical profession's expense. Thus, interprofessional relations in the health care hierarchy are portrayed as inherently conflictual, a reflection of the dynamics of a zero-sum game in which junior occupations attempt to undermine medicine's dominance by carving out independent domains of practice (see Freidson 1973b; Larson 1977; Child and Fulk 1982; Abbott 1988). However, this assumption cannot bear the weight of empirical evidence easily. The EMTs I observed readily acknowledged and seemed to accept their subordinate status within the health care hierarchy. The formal authority and expertise of physicians were never questioned, nor did EMTs claim that their work was substantively different from that performed by emergency room staff members. In fact, EMTs fondly referred to their ambulances as "mobile emergency rooms" and boasted of being able to perform the same tasks as an emergency room (ER) staff during the first 30 minutes of patient contact—hardly a promising basis for launching a campaign for power and independence.

Nevertheless, EMTs did object to the notion that they possessed no skill or knowledge independent of the physicians'. While they applied the same skills used by other health care practitioners, they did so in a vastly different context. Away from the sterile, safe, and relatively controlled environment enjoyed by physicians and nurses, the EMTs conducted their work in settings which were often dirty, crowded, noisy, and downright dangerous. "I think there's more involved . . . in practicing in the field than in the ER," insisted one EMT. "A lot of times we're working under conditions we can't control. We could have a call during a thunderstorm where I'll start an I.V. in the rain and lightning. Or I'll have to crawl into a wreck to intubate a patient who's still trapped in a car." An operating room nurse who also worked as a part time EMT concurred:

> The basic things that you take care of—airway, breathing, circulation—never change no matter where you are. But doing things in the field is different because you don't have optimum circumstances. If I'm in the hospital and I intubate a patient, he's already sedated and had suscitocholine [to suppress a gag reflex] and

> his stomach is empty. When you're out in the field you don't have this luxury . . . [so] the patient will be gagging and puking while you're intubating him, and it's really hard to do. You also have to deal with hysterical families screaming and yelling at you, or trying to start an I.V. on somebody pinned in a car. I'm telling you it's tough, and the guys who deal with this every day are very, very good.

In short, practicing in this chaotic environment necessitated elements of speed, versatility, improvisation, physical and emotional control, and interpersonal finesse which distinguished the practice of EMTs from that of hospital-bound practitioners. As a result, the EMTs' skills and abilities were complementary rather than redundant.

EMTs claimed that after repeated displays of technical prowess and responsible decision making, some physicians and nurses gradually came to recognize their expertise and would thus defer to their judgment in matters pertaining to patient care in the field. This was described as building a "trusting" or collaborative relationship with other health care practitioners. Collaborative relations were prized for two reasons. First, they afforded EMTs a degree of autonomy which belied their hierarchical status. As one EMT explained, "The first few times you work on an ambulance squad, the doctor really looks at what you're doing. They'll ask for a cardiac strip to be sent [to the ER via telemetry] to check out your interpretation . . . and give orders for everything. But after three or four good interpretations, the doctor will start trusting your interpretations and stop telling you what to do." Second, establishing collaborative relations with other health care practitioners recognized and reinforced the EMTs' membership within a local community of practice—a structure which afforded them a degree of participative equality or status nonexistent in their role as physician extenders. For example, in cases where an EMT believed that the mandated treatments were ill advised or that improper orders had been issued by medical control, they negotiated with the physician. "Out here we're the doctor's eyes," observed one EMT. "Because you're actually looking at the patient you can tell if they don't need something. Usually you can tell the hospital that and they won't argue."

EMTs were quick to note that they were not alone in benefiting from a more collaborative arrangement. One EMT claimed that because their skills were complementary to those of hospital-based practitioners, the EMT, nurse, and physician approximated a tripartite system not unlike that comprised of the president, Congress, and the Supreme Court. Just as each branch of government provides a check and balance to the power of the others, so the EMTs' partnership provided a similar check on the judgment and autonomy of the hospital-based practitioners, thus increasing the odds that decisions regarding patient care would be optimal.

The EMTs' role as partners in the emergency care process provided the canvas on which the normative standard of care, a set of rules fashioned by EMTs themselves in concert with other members of the local community of practice to guide everyday work, was drawn. This was "the spirit of EMS," a code of behavior superimposed on the mandated standard of care embodied in EMS protocols. The spirit of EMS did not amend the technical content of protocols. Recall that EMTs respected the physicians' technical expertise and never claimed that they had a more expansive or detailed knowledge of anatomy, pharmacology, and related disciplines; to question the purely clinical aspects of a protocol penned by physicians was therefore considered inappropriate. Rather, the spirit of EMS called for carrying out protocols in a manner which displayed an intense concern for the patient's welfare.

This concern was demonstrated in three ways. The first consisted of proactivity: drawing on one's experience and knowledge to modify practice to achieve optimal care, and taking the initiative in requesting orders or administering treatments. Although a modest amount of hesitation and strict adherence to protocols—or "going by the book"—was understandable and even advisable for novices lacking field experience, it was expected that such behavior would give way to confidence and improvisation as the EMTs gradually perfected their craft. Indeed, that trainees were exhorted to adopt a more proactive role in the medical encounter even before active field practice was seen when an EMS instructor urged her charges to ". . .ask the doctor for [permission to administer] a drug instead of waiting for them to order it. I want you to intercede on the patient's behalf." The spirit of EMS was also evident in a willingness to disregard personal comfort and ease in the interest of aiding patients. For instance, on one occasion an EMT displayed a hand badly swollen from multiple wasp stings sustained while rappelling down a steep embankment to reach an injured hiker. When asked if a less hazardous way of reaching the patient had been available, the EMT allowed that there had, but that this option would have delayed the arrival of help considerably. He added that the discomfort—which was not inconsiderable—was a small price to pay for a successful rescue. A final means of demonstrating the spirit of EMS was by showing empathy and compassion for patients and their families, and respecting their wishes whenever possible. "In ambulance work," noted one EMT, "you can walk in, slap oxygen on somebody, and load them and go. You don't have to have . . . empathy for the people you transport. But that's pretty lousy care. Good [EMTs] know that EMS is a caring profession."

The EMTs' normative standard of care also specified whether certain types of action would occur during emergencies. The mandated standard of care dictated that every patient be treated alike and receive virtually the same quality of care. EMTs insisted that every patient did receive the same care from a purely clinical standpoint—no one was ever denied medical attention or given substandard care and transport. However, not everyone merited care given in the spirit of EMS, which was reserved for cases judged to be "true emergencies." To understand their selectivity, one must realize that the EMTs, particularly those employed by Medco and Jones, were frequent victims of "system abuse." Basically, system abuse entailed summoning an ambulance when a genuine medical emergency did not exist. According to the crews, system abuse was motivated by several factors. Often a would-be patient might want a ride to the hospital to visit friends or seek treatment for a minor or chronic illness but be unwilling to part with taxi or bus fare. Homeless people would feign illness to gain entry to the hospital and thus pass five or six hours in a warm, dry place; some called for an ambulance simply to relieve their boredom and loneliness for a few short hours.

Without exception, EMTs resented this behavior because it seemingly denigrated their expertise by casting them in the role of "taxi drivers," rather than emergency care specialists. System abuse also represented a tremendous waste of equipment and manpower, a drain which jeopardized the overall quality of care the squads could otherwise provide. When asked about the consequences of abuse, EMTs like this Medco employee exploded with anger: "You want to know about system abuse? Well, a couple of weeks ago we were called to a house at four in the morning to administer a Band-Aid. Can you believe it? A guy nicked his finger and didn't have a Band-Aid in the house! Meanwhile, while I'm giving this asshole his Band-Aid, a code [cardiac arrest] came in . . . Because there are only three crews on duty after midnight, a rig had to come across town to answer the damn code! That guy was long gone by the time they got there." In this case, the delay indirectly caused by a trivial, "abusive" call apparently played a role in the second patient's death, a situation which this EMT found nearly intolerable.

EMTs responded to this frustrating behavior in the only way they could: by withholding the spirit of EMS. As legal agents of the physician, they were obliged to adhere to protocols and assist system abusers by rendering any care or treatment requested, but they did not need to render it in compliance with their normative code. Hence, rather than evincing compassion or concern, they gave treatment in a sullen and perfunctory manner. The EMTs were unfailingly polite, but little attempt was made to communicate with "patients" or their families beyond what was necessary to inquire for a brief history and description of circumstances surrounding the call. Unlike ambulatory patients who presented some genuine medical problem, system abusers were never helped into the ambulance by a solicitous crew member and, during transport, crews neither made eye contact with the "patient" nor engaged them in the small talk used to allay fear and elicit medical clues. Nor were crews willing to sacrifice their personal comfort in obvious cases of system abuse. For instance, protocols specified that all patients, including system abusers, were to be carried to an ambulance on a stretcher. But EMTs considered system abusers undeserving of their toil. "If I carried every one of these jerks," protested one EMT, "my back wouldn't last a week! There's nothing wrong with them, so they can walk to the rig."

In short, the normative standard of care specified how and when EMTs performed their work. It also established their social obligations and responsibilities. As physician extenders, the EMTs' duty was clear: to give good patient care, one adhered scrupulously to protocols. The EMTs' duty as partners in the emergency care process was also quite clear: to give good patient care, one followed the dictates of experience and conscience, even if these contradicted mandated standards. Basically, this implied doing more or less than protocols demanded if the EMT believed that the patient would benefit from the deviation. One EMT summed this up nicely when he flatly stated, "We do whatever it takes to give the best patient care possible. Period."

For their part, EMTs argued that some protocols were wholly impractical, such as a rule which required that patients be treated where they lay rather than within the safe, clean confines of the ambulance. Any gains in speed made by offering such treatment would be more than offset by the distractions caused by inclement weather and unruly bystanders as well as the increased risk of infection incurred by providing treatment in a polluted environment. Similarly, some protocols were considered trivial, like those which required repeated assessments on patients who were essentially dead. But perhaps most onerous were those protocols thought of as being needlessly cruel. The rules regarding resuscitation in cases of sudden infant death syndrome (SIDS) provided one such example. Because of their limited training, EMTs were not permitted to pronounce someone dead and thus withhold treatment unless the victim was "obviously dead"—decapitated or displaying signs of advanced rigor mortis or decomposition. But because victims of SIDS rarely displayed such signs, EMTs were obligated to begin resuscitation even when it was futile, a move which cruelly raised the parents' hopes that their child might be saved. In all these instances, deviations from protocols were considered to be both justifiable and honorable.

### *The Mandated and Normative Standards Considered*

In summary, we can see that the EMTs' moral consciousnesses were shaped by both mandated and normative standards of care. According to the former, standards of care were encoded into a series of protocols. To give good care, EMTs were expected to adhere to these protocols and the direction of medical control. They might voice concerns about the benefits of a given treatment, but were nevertheless legally obligated to follow the dictates of medical authorities. But according to the normative standard, providing good care implied rendering protocols in the spirit of EMS. This standard of care also prescribed various standards and rules to guide the EMTs' on-scene action and behavior but, unlike protocols, normative rules concerned not the purely clinical content of tasks, but rather how and when certain tasks were to be done. To give good care, EMTs were expected to draw on their experience to modify practice, put the patient's comfort ahead of their own, and show empathy and compassion whenever patients presented a genuine need for medical care and transport. They were obliged to "do whatever it took" to ensure the patients' welfare—even if this meant deviating from mandated rules from time to time.

There are, of course, some notable discrepancies and points of tension between these standards. Under certain conditions, the tension sparked by these discrepancies flared into moral conflicts which represented a great source of difficulty. In the remainder of this chapter, I will examine situations where this occurred and the ways in which EMTs responded to these conflicts in the context of everyday work practice.

The observation that dual membership in institutions embracing different forms of social organization gives rise to moral conflict is hardly novel. Sociologists of work and occupations have often remarked on the "strain" or "tension" which results from employing horizontally ordered occupations like crafts and professions in vertically ordered institutions and organizations (see Kornhauser 1962; Pelz and Andrews 1966; Raelin 1985; Von Glinow 1988). Similarly, ethicists have assembled a voluminous literature on the moral conflicts—variously known as "whistle blowing," "double agent" dilemmas, "divided loyalty" dilemmas, and dilemmas of "organizational disobedience"—which presumably arise from this situation (see Institute of Society, Ethics, and The Life Sciences 1978; Center for the Study of Ethics in the Professions 1982; Murray 1986; Toulmin 1986). However, in doing so, both groups tend to exaggerate both the prevalence and the severity of this conflict. For example, Zabusky (1993) has noted that sociologists have long considered the demands made by vertical and horizontal work systems to be virtually incompatible; actors subject to both are therefore thought to become "dysfunctional" employees prone to a variety of "maladaptive" behaviors. The fact that ethicists have elevated these conflicts into dilemmas—quandaries which necessitate choosing between equally unpleasant alternatives—suggests that these conflicts are viewed as extreme. One is left with the impression that social scientists consider the juxtaposition of vertical and horizontal work systems to lead inevitably to dissonance and its unhappy outcomes—discord, malfeasance, and attrition.

Yet empirical evidence drawn from my study of EMTs suggests that social actors are far more skillful at negotiating this conflict than this literature seems to convey. In fact, the EMTs I studied regularly weighed the implications of these conflicts, and only occasionally would the behavioral outcomes of their deliberations match those typically depicted by sociologists and ethicists. Consequently, relatively few of the conflicts which abounded in their everyday work life were actually perceived as dilemmas. Hence, it is inappropriate to assume that the tension between vertical and horizontal work systems necessarily produces conflict, or that moral conflicts, when they do occur, escalate to the proportions of true dilemmas. Instead, one must ask about the methods used to manage these tensions and under what conditions these tactics become untenable, for only by asking these questions can we fully understand the dynamics and implications of this moral mismatch. With this point in mind, I now turn an explication of the EMTs' approach to managing moral conflict.

## ON CONFLICT AND COMPROMISE: RESPONSES TO DISPARATE STANDARDS OF CARE

The EMTs' responses to tensions spawned by clashes between mandated and normative standards of care were derived using a moral calculus that weighed possible courses of action according to the perceived risk to the patient versus the perceived risk to themselves. This calculus, which was applied whenever EMTs believed a conflict existed between the dictates of the mandated and normative standards of care, can be diagrammed using the two-dimensional framework seen in Figure 7.3. The first dimension, perceived risk to patient, relates to the issue of the patient's health and well-being. Patients were classified as "low risks" if their illnesses and injuries were not severe, or if they were not ill at all, as in cases of system abuse. Because these afflictions were not life threatening, there was very little chance that the patient's condition would "go downhill fast" or deteriorate rapidly during pre-hospital care and transport. Patients were also considered "low risk" if they were very close to death and medical intervention was unlikely to make a difference. When patients were labeled as "high risk," however, medical intervention could make the difference between life and death, or between a good or poor quality of life after the incident. These patients were afflicted with serious illnesses and injuries and were invariably in need of prompt medical attention.

The second dimension of the framework describes the EMTs' perceived risk to self; it reflects the nature and severity of sanctions associated with the strategies used by EMTs to manage moral tension and conflict. Incidents were classi-

| | | Risk to patient | |
|---|---|---|---|
| | | Low | High |
| Risk to self | Low | Negotiation | Willing compliance |
| | High | Reluctant compliance | Variable response |

FIGURE 7.3
EMTs' behavioral response to perceived conflicts of interest

fied "low risk" if a deviation from mandated rules was unlikely to be discovered by medical authorities, or if EMTs expected to receive little or no reproval upon being caught. In turn, these outcomes hinged upon two factors. The first was the EMTs' relations with medical control. If the physicians and, in some circumstances, nurses who comprised medical control were members of the EMTs' local community of practice and a bond of trust existed between them, "reasonable" deviations from protocols would often be intentionally overlooked. The second factor concerned the nature of the incident itself. Although protocols existed for practically all medical conditions and other on-scene difficulties, it was impossible to anticipate every contingency which might arise during an emergency. Consequently, there were some "gray areas" in the protocols, or areas in which the particulars of on-scene behavior remained unspecified. If detected, deviations occurring within the context of an acknowledged "gray area" were open to some negotiation or debate. Conversely, incidents were considered to be "high risk" if any deviation from the rules was likely to be discovered and punished. These incidents did not fall into a "gray area"; they concerned cases in which the EMTs' roles and responsibilities were clearly defined within the protocols and for which a precedent of punishment was firmly established for deviations. These conditions made it difficult for sympathetic physicians and nurses to overlook departures from protocol.

### *Response No. 1: Negotiation*

In cases where the risk to the patient and the risk to themselves was believed to be low, EMTs negotiated conflicts by engaging in minor rule-breaking. The logic behind this strategy was simple—if negative consequences were unlikely to occur, the easiest way to contend with a problematic point of protocol was to ignore it. This often took the form of surreptitious acts such as giving mildly exaggerated radio reports to receive desired orders or recording multiple sets of vital signs for dead patients when in fact only one set was taken.

This behavior occasionally assumed a deliberate, game-like flavor. This was witnessed during a call in which a crew was summoned to help an elderly man who was having difficulty breathing. Although the patient was not in immediate danger, the crew decided his condition did merit a trip to the hospital. There was, however, a problem: the patient refused to accompany the crew unless transported to a hospital that was temporarily "closed." Hospitals in the local emergency care system were closed when their emergency departments became clogged with patients. Such facilities would not accept emergency admissions for several hours, and radio requests for admission under such circumstances were routinely denied. Ambulance crews would therefore be forced to take patients to other care facilities. Because this patient refused to consider any other facility, however, the EMTs were faced with the choice of abandoning a patient whose condition might conceivably worsen or disregarding protocol by simply transporting him to the hospital unannounced, a move guaranteed to anger the emergency room staff. In this case, the crew gleefully chose the latter course of action:

> Bill called over to me and said, "Wait and see. The nurse will bitch me out because we never called in this guy." Just as predicted, we were met outside by the triage nurse. She began to loudly harangue the crew for not notifying the ER before bringing in a patient. Bill shot back, "What am I supposed to do? It's the dispatcher's job, not mine. Go complain to him." He turned on his heel and returned to the rig, leaving the speechless but fuming nurse at the curb. Back at the rig, Bill began to chuckle, and then laugh. He raised the pitch of his voice to imitate the nurse in a squeaky, mocking tone: "Oh, you didn't call us, you didn't call us before coming . . ." We both erupted in laughter. Bill then announced that the next drunk they picked up would be brought here. In fact, every drunk they picked up tonight would be brought back here to enjoy the nurse's company. As his laughter subsided, Bill remarked, "It's all a game anyway." "Oh? What's the object?" I asked. Bill grinned and said, "To see just how much we can piss each other off." "And who sets the rules?" "I do. That's why it's such a good game."

The crew remained true to its word; that entire night every troublesome, obnoxious patient was taken to this same hospital, even if other facilities were closer.

To understand the EMTs' admittedly spiteful behavior, it is important to note that while other members of the community of practice didn't have to agree with the EMTs' action, they were expected to at least listen to the EMTs' rationale for acting as they did. This expectation presumed that medical staffs recognized that protocols were occasionally out of touch with the practical exigencies of the EMTs' work; that extenuating circumstances made deviations understandable, if not acceptable, in these cases; and that EMTs had the technical and moral ability to undertake such deviations responsibly. The nurse's outburst contrasted sharply with the respectful give-and-take that marked negotiations when these presumptions were tacitly accepted by all members of the community of practice. "If we have a disagreement with the hospital during a call," explained one EMT, "we'll sit down and talk with them afterwards at the ER . . . We'll describe the call and say, 'This is why I thought the patient needed something or didn't need something.' If they trust you, they'll usually agree." In other words, EMTs attempted to reconstruct the context in which the deviation occurred for hospital-bound practitioners. Only when the logic dictating their actions was placed in context could the EMTs obtain the footing necessary to persuade others that their decision had in fact been the proper one. To do this, however, they needed to have the opportunity to explain their actions. The triage nurse's outburst had proven objectionable precisely because it curtailed this normative process. The EMTs' behavior therefore served as a reminder that, as a member of the local community of practice, the triage nurse was expected to understand and comply with these norms.

*Response No. 2: Willing Compliance*

Because their normative code stipulated that EMTs had a duty to provide the best possible care for patients, EMTs generally complied with protocols when the perceived risk to the patient was high but the perceived risk to themselves was low. At first glance, this behavior seems contradictory—EMTs did, after all, boldly vow to "do whatever it takes" to provide good care, and this usually implied deviating from protocol. But sometimes doing whatever it took required deviation from normative rules as well. The normative standard of care specified that the EMTs' foremost responsibility was to the patient. Hence, even when protocols were considered trivial or impractical—essentially constituting an annoyance for the practitioner—EMTs were expected to "bite the bullet" and resist the temptation to second-guess physicians by taking "shortcuts" if such action risked affecting the patient's condition adversely. To do less displayed a callousness and lack of humility which EMTs considered reprehensible.

The use of procedural shortcuts provides a good example of this expectation. Shortcuts consisted of eliminating, collapsing, or rearranging certain steps in a treatment protocol. EMTs argued that such actions were justified because experience had taught them that some steps were unnecessary or counterproductive. This practice was, however, guided by strict norms about who could take shortcuts and when they were appropriate. For example, neophytes were not expected to take shortcuts; such measures were introduced to them gradually, under the tutelage of more experienced EMTs, after the new EMTs had spent several months "in the field." Even accomplished EMTs were expected to refrain from the practice when the patient's condition appeared serious. This was demonstrated during the case of an "MVA," or motor vehicle accident, wherein an EMT took a shortcut despite the questionable status of his patient:

> After finishing his initial assessment and noting that the patient seemed reasonably stable, Stan, the crew chief, announced that they wouldn't administer ALS (advanced life support)—after all, they were scarcely a mile from the ER. But the other members of the crew didn't agree. As he was recording the vitals, Dan repeatedly asked Stan if he didn't want to take a closer look at the condition of the car—it was badly twisted. Pam chimed in and noted that bystanders reported that the patient had been unconscious when pulled from the wreckage. Both were broad hints that ALS was indicated in this case. With an annoyed and impatient

> look, Stan reminded them that he had already seen the vehicle. They were leaving and the patient wouldn't receive ALS. After the call, Dan and Pam confronted Stan directly. The patient [had] clearly shown potential for "going downhill" or failing fast. This was no time to be taking shortcuts. Stan plaintively claimed that the doctor had voiced no complaint about his decision, but his peers were unmoved. Both insisted that Stan had shown questionable judgment . . .

The crew chief's action proved objectionable for two reasons. First, interactions between experienced EMTs were subject to the same norms of negotiation and debate as existed with hospital-bound members of the community of practice. Hence, refusal to consider his peers' opinions betrayed a dictatorial style which was distinctly non-collaborative and hence unacceptable. But more disturbing was the apparent disregard which the chief displayed for his patient's welfare. Although the supervising physician did not reprimand Stan for taking an obvious shortcut—a move which signaled tacit approval for the deviation from protocol—the other crew members considered his behavior to be both lazy and irresponsible. Indeed, in the weeks following this incident, Stan's behavior made him the butt of numerous jokes which pointedly questioned his attitude and judgment ("So, Stan, I hear you're a doctor now . . .") or his dedication to patient and craft ("Stan, hand me that piece of tape—if it isn't too much fo you . . .") The implication of this incident is that, when the patient's welfare was at stake, EMTs were expected to temporarily subordinate any dislike of protocol and trust the technical competence of the physicians who had drafted the procedures.

### *Response No. 3: Reluctant Compliance*

In cases where the risk to themselves was judged to be considerably greater than the risk to patients, the EMTs' instinct for self-preservation won out over personal convictions regarding the necessity or appropriateness of certain points of protocol. Because deviations from protocol were likely to be detected and punished, perhaps severely, and because patients would likely suffer few ills from receiving the protocol in toto, EMTs typically opted for the safest course of action: compliance with mandated rules and standards.

Although this response basically entails the same type of action discussed in the preceeding section, the EMTs' attitude regarding the strategy of compliance in these instances differs sharply. When the perceived risk to the patient was high, EMTs had a normative duty to set aside personal dislike of protocol. Thus, compliance with protocol represented the most "honorable" course of action. However, this was not the case when the risks were weighted in the opposite way. The only reason EMTs complied in these instances was because they feared the personal consequences of failing to do so. Convictions about the lack of necessity or inappropriateness of protocol were not suspended. Because this action basically entailed a violation of the EMTs' normative rules regarding good patient care, EMTs complied with protocols only with considerable reluctance and regret. The dynamics of this response were described by an EMT who had recently made such a decision:

> We were rehashing today's calls and got on the topic of resuscitation in cases of imminent death. Jim shook his head and said, "There is no dignity in an emergency death." He explained that while there was no greater thrill than a "save"—basically reviving a patient near death—doing so on a terminally ill patient seemed pointless and cruel. He therefore allowed dying patients dignity by sparing them from the more gruesome aspects of resuscitation whenever possible. For example, he recently had a case in which the family of an elderly stroke victim summoned the crew to take their dying mother to the hospital. The family asked the crew to withhold ALS and he wanted to respect their request, but they had no DNR [Do Not Resuscitate order]—an absolute necessity in these cases. The hospital did order ALS in route, and Jim complied. The patient died shortly thereafter. Jim admitted that this case troubled him deeply: "These decisions create a lot of stress for us," he said. "I'd like to follow the family's wishes, because the patient's quality of life is an important consideration. What good does it do to have Mom vegetate in the ICU [intensive care unit]? But I have to cover my ass, too."

This response was prompted by two factors. First, as the EMT indicated, this call did not fall into a "gray area"; there is a well-established legal precedent requiring possession of a DNR order if health care practitioners are to withhold care. Even so, EMTs admitted that it was possible to give a radio report of the patient's condition in such a way that the hospital staff would refrain from ordering advanced life support. This option was precluded in this case, however, by the absence of a sympathetic physician in medical control. The physician on duty at that particular time was well-known for being a "stickler for protocol"—unfavorably disposed to granting EMTs a wide degree of latitude. Hence, this EMT believed he had little choice but to disregard the patient's wishes.

The inability to provide even a small comfort to the patient's family was a violation of normative code which bit deeply into this EMT's moral consciousness. But these cases were also disturbing because they heightened awareness of the EMTs' subordinate position in the health care system. The normative rules fashioned within the community of practice are informal and situational; that is, they exist only with the indulgence of sympathetic physicians and other practitioners in medical control. Outside this small social circle, EMTs have no privilege or status other than that bestowed upon them by their formal role of physician extender. Accordingly, their ability to exercise their normative ethos and work practices is highly circumscribed. Compliance with protocols in such cases therefore served as a frustrating reminder of the EMTs' anomalous status within the health care system.

*Response No. 4: Variable*

When both the perceived risks to patient and self were high, the EMTs were confronted with a problem which offered no obvious solution. This occurred whenever the EMTs believed that the standard of care specified in the protocols was downright harmful to the patient but failure to comply with protocol would be harmful to themselves. The conflict between the standards of care therefore escalated to the proportions of a true moral dilemma, forcing the EMT to choose between equally disagreeable alternatives: following protocol when he or she believed that mandated rules were counterproductive, violated normative standards, and could possibly inflict harm on the patient; or following normative rules in violation of protocol, which invited charges of malpractice, fines, loss of certification, and even imprisonment. Due to the urgency of the situation, even indecision represented an untenable option. EMTs in this unenviable position were forced to choose literally between doing what was expected of them as physician extenders and doing what they believed their work demanded.

Responses to these dilemmas were variable, guided by individual dictates of conscience. For instance, after a quick but careful estimate of the situation, some EMTs complied with the dictates of protocol and consequently suffered the personal and social consequences of not doing what they truly believed to be in the patient's best interest. This choice and its painful aftermath is illustrated in the tale recounted by an EMT who once made this difficult decision:

> We now do fairly advanced procedures for airway management. We can intubate a patient . . . to breathe for them, and use forceps to remove small objects. But unfortunately, when they were started, [these procedures] were authorized for adult patients only . . . We had a pediatric call where a two-year-old kid had a stone caught in his throat and was choking. We had no air exchange at all. We had all the equipment and if it was an adult we could have intervened but, being that we didn't have treatment protocols for pediatrics—you really felt useless, you know? We were so far out . . . the kid is turning blue, and then purple, and then he died. There was nothing we could do. We knew what was going on, we knew what was in there, and we knew what we needed to do, but we couldn't do anything about it. Those are the ones that really hit home.

Although this incident had occurred several years before, the EMT was still haunted by the feelings of frustration and impotence he experienced during this episode. The social consequences of this response were far more subtle but no less salient. The EMT admitted that no one had openly sanctioned his failure to act; it was, after all, only a matter of time before they themselves would face the same situation, and would likely act similarly. Yet, he also suspected that his peers' silence masked a contempt for failing to uphold normative standards: "Nobody pats you on the back for going by the book."

In contrast, others who faced these dilemmas chose to ignore protocol and thus risk incurring the legal and financial consequences of this course of action. Such a decision was a sobering one, for the penalties levied against such flagrant disregard of protocol meant that the patient could be the EMT's last. This was the case for one crew who made such a decision:

> There was a call for an MVA out on Route 20 where the patient hit the steering wheel and crushed his trachea. There was absolutely no air exchange. The crew tried all the usual procedures—intubation, oxygen. Nothing worked . . . So, as a last resort—and it was a last resort, [the crew] did a tracheotomy. Problem was, this procedure hadn't been officially accepted at [the] regional level. The crew was brought up on charges before [regional authorities], even though they had exhausted all their resources and the patient was dying in front of them. They had a procedure they'd been taught to use, gone through a clinical experience in training, had the equipment to do it, and did it only as a last resort. Yet they lost their EMT cards because of it . . . [The review board] said that all their experience and classroom training wasn't "official," that they didn't "officially" know how to do it so they shouldn't have tried to save the guy's life. Which was garbage, you know? We all left the meeting saying, "If that accident happens again I'm going to do the same damn thing . . ."

After the incident, the crew's defiance catapulted them into the realm of legend within the local EMS community. Although this incident had also occurred several years before, EMTs in the area continued to speak approvingly of the crew's decision. Yet, there was a darker side to this celebration, for they also admitted that the crew's victory had been a Pyrrhic one which few would readily imitate. In the end, EMTs knew that, for all their collaboration and concerted rule-breaking, they remained institutional subordinates and were ultimately subject to the strictures and limitations inherent in that status.

## DISCUSSION AND IMPLICATIONS

The strategies of rule-breaking and ready compliance seen when risk to self was low were useful in ameliorating tensions between standards of care. But when risk to self was high, EMTs were far less successful in coping with these tensions. The key to understanding this difference lies in the conditions which distinguish a situation as low or high risk in the first place.

Situations were perceived as low risk precisely because formal, hierarchical roles and rules were suspended in favor of the norms of the local community of practice. This was made possible by the existence of sympathetic authority figures in medical control—physicians and nurses who understood that the EMTs' contextual skills and moral character were complementary to their own. Typically, these practitioners also had experience "in the field," which allowed them to witness firsthand the EMTs' technical prowess and judgment (see Pepe and Bonnin 1989). Apparently recognizing the practical benefits of allowing EMTs to exercise their judgment and contextual expertise in a more liberal manner, these practitioners willingly established a collaborative working relationship with EMTs in which disagreements were settled by respectful negotiation rather than by the exercise of formal authority. In short, despite their hierarchical position within the medical division of labor, some medical authorities subscribed to a moral vision consonant with the EMTs' own. This collaboration made it possible for EMTs to engage successfully in minor rule-breaking without penalty, just as the sense of mutual respect or "trust" fostered within the community made it possible for EMTs to subordinate their opinions in cases where being wrong could be costly without loss of face.

In contrast, some incidents were perceived as carrying high personal risk because the formal roles and rules characteristic of the health care hierarchy prevailed. The community of practice was informal and situational; its existence hinged on the presence of cooperative, like-minded figures in medical authority who were willing to bet their medical licenses on the EMTs' technical and moral competence. Because of an unfamiliarity with the EMTs' work and a general wariness of their technical and moral qualifications, many physicians and nurses were unwilling to grant EMTs an expanded role in the patient care process. This was most obvious among practitioners who had little contact with EMTs and who, consequently, had scant knowledge of their contextual expertise or moral character. Lacking such exposure, practitioners based their understanding and actions upon cultural images which portrayed EMTs as subordinates with limited skills and questionable character.

But even physicians and nurses convinced of the EMTs' technical and moral competence were themselves answerable to authorities who took a dim view of allowing EMTs an expanded role in the emergency care process. Because the architects of EMS—historically, federal legislators and, more recently, state, regional, and local administrators—have subscribed to a moral vision projecting only a limited role for EMTs, they have built an institution which does little to recognize or foster the normative systems through which EMS work is actually accomplished. When the physicians' decision to allow the EMTs latitude was likely to be questioned by others, it became difficult to suspend formal roles and rules. Thus, physicians had little choice but to assume their formal roles as medical authorities and enforce protocols. This, in turn, forced a hardship upon EMTs: deprived of the status and prerogatives of their normative role, they had little choice but to follow the dictates of protocol or to disobey and accept the legal and financial consequences of honoring normative rules. Both options served as a frustrating reminder that, from an institutional standpoint, EMTs are functionaries with no authority or status aside from that formally conferred upon them by their association with the medical profession. This glaring contradiction between cultural image and self-conception reportedly represented a source of confusion, resentment, and, eventually, attrition.

The EMTs' experience provides an excellent example of the practical and moral consequences of failing to entertain the notion of moral plurality. As vertically ordered work systems become increasingly horizontal or "occupationalized" (Barley 1991), the presumption that a single moral vision holds hegemony becomes increasingly tenuous. The expansion of technical work and the occupationalization of firms gives rise to a plurality of moral visions, with each playing a pivotal role in shaping cultural images. Wittingly or unwittingly, employers and policy makers who fail to recognize and accommodate moral visions other than their own risk drafting rules and policies which can seriously impair the effective and efficient operation of the normative systems through which work is conducted. The EMTs' experience illustrates the fact that not only technical workers are thus affected; the general public may also suffer if work systems do not allow workers to do what they know how to do. Moreover, because the moral visions characteristic of vertical and horizontal modes of organization impose different, and somewhat contentious, roles and obligations, their juxtaposition creates a potential for conflict and dissonance. This conflict extends beyond the occasional overt crisis such as occur when EMTs must choose between risks to patients and risks to themselves—these occasions merely serve as reminders of the inconsistencies of status and social identity which technical workers suffer unresolved.

These problems and their outcomes are real and undeniably serious. Yet it is hardly surprising that the notion of moral plurality has met with neglect or resistance, for the mores embraced by technical workers directly challenge a number of fundamental assumptions about authority and employment relations in hierarchical work settings. For instance, the very notion of moral plurality strikes down the idea that there is "one best way" for accomplishing any given task, and that it is management's duty to chart this course for the collective. Similarly, the norm of negotiation and debate inherent in horizontal forms of social organization contradicts expectations of obedience to managerial dictates in the interest of achieving selected ends. Perhaps the greatest challenge is given to the disparate images of technical and moral fitness within the hierarchical division of labor. In vertically ordered settings, the ability to command legitimately is grounded in the presumption that superiors possess technical and moral qualifications which their subordinates lack. However, in horizontally ordered work settings, workers are presumed to have similar technical and moral competencies (provided they have achieved full membership). The presumption of similarity robs hierarchical authority of its legitimacy by providing actors with comparable degrees of technical and moral footing. From this standpoint, management possesses no authority apart from that wrangled from other workers by virtue of their substantive expertise and persuasive abilities.

The implication of this assessment is clear: to design work systems more congenial to the technicians' mores and thus circumvent problems associated with neglecting moral plurality, employers and policy makers must be willing to radically rethink their views of institutional morality. Absent this, it seems unlikely that any alterations made in work systems will hold. This situation invites further exploration of both the technical and moral dimensions of authority, particularly assumptions about the nature and distribution of knowledge and honor within the social division of labor. Of special interest is the cultural tendency to value formal, theoretical knowledge over the contextual knowledge derived from everyday practice. A growing body of sociological literature suggests that this preference is based on factors other than utility (see Brown and Duguid 1991; Barley 1993). Perhaps the relationship between formal knowledge and authority holds the key to understanding this preference.

Also required is an understanding of how social institutions shape and replicate these assumptions. The fact that formal knowledge and reason are privileged over experiential knowledge and judgment in the vertically ordered workplace provides a powerful argument for more detailed examination of education's role in this process in general, and the social significance of credentials in particular. The capacity of educational credentials to signal possession of certain types of skill and knowledge is well established, but their role in the attainment of honor is less frequently examined. In the distant past, a heightened sense of moral character was assumed to be an outcome of noble birth. The spread of Christianity and monarchical power gradually transformed the meaning of honor from a characteristic of noble blood to a mark of fidelity and devoutness (Haber 1991). Status and authority in the division of labor became symbolic of favored positions in the eyes of God and king—omniscient and divine entities who presumably lent their favor with utmost discretion. But in our secularized age of reason, spiritual sources of legitimacy have given ground to formal education, particularly in the sciences (Midgley 1992). The ability to discern socially advantageous means and ends is now secured by the powers of rationality and insight, presumably imparted through formal education. Thus, the diplomas and certificates granted by colleges and universities have become prized for their ability to signal the attainment of honor as well as skill (Weber 1968; Collins 1979), and credentials have become a pivotal means by which society takes stock of individual character.

That examinations of the social significance of credentials must ultimately extend beyond the classroom and workplace to the larger institutions comprising the very framework of society is amply demonstrated by the research I have described. For instance, the EMTs' experience suggests that credentialism has permeated the structure of regulation quite thoroughly. Although experienced EMTs possessed a wealth of contextual skills and knowledge about health care in prehospital settings, their lack of the credentials typically deemed necessary for a more active and responsible role in patient care relegated them to a legal status in which medical activities and responsibilities were highly circumscribed. From its inception, the occupation's authority to care for the sick and injured has hinged entirely on their formal (read subordinate) relationship with physicians.[6] In fact, if divorced from their relationship with physicians, EMTs possess no medicolegal rights or protections other than those enjoyed by any average citizen without medical training. This observation suggests that credentials have become a powerful social proxy for deeply rooted assumptions about the technical and moral fitness of certain groups within the legal and regulatory establishment.

Moreover, the pervasiveness of these assumptions directly impedes the formation of the collaborative work systems characteristic of horizontal forms of social organization. Because EMTs are merely legal agents of the physician, physicians must bear ultimate responsibility for the EMTs' actions; hence, even though they might personally be convinced of the efficacy of more collaborative work systems, it was in the physicians' interest to enforce protocol and thus limit any potential liability to themselves. This is particularly true given the increasingly litigious nature of Western society; the everpresent specter of lawsuits has made even sympathetic physicians understandably reluctant to allow EMTs a more active role in the care process. This picture demonstrates that a failure to keep pace with the technological and social shifts that have enabled the EMTs' evolution from "delivery persons" to emergency care specialists has also transformed regulations intended to protect the public into a social liability.

6. In fact, Boyd (1983) reports that Congress considered the EMTs' medicolegal status as physician extender to be a "non-negotiable" item in its formulation of the Emergency Medical Services System Act of 1973.

# PART III

# Implications of Technical Practice for Training, Credentialling, and Careers

# 8

# THE INFAMOUS "LAB ERROR": EDUCATION, SKILL, AND QUALITY IN MEDICAL TECHNICIANS' WORK

*Mario Scarselletta*

## INTRODUCTION: SKILL IN TECHNICAL WORK

What is the nature of skill in technical work and technical occupations? What are the skills that technical workers must possess in order to perform their jobs well? Where and how do technicians acquire these skills? As technical occupations continue their ascent as the fastest-growing segment of the U.S. economy (Barley 1991), employers and policy makers inevitably find themselves wrestling with these questions. Questions relating to skill are at the heart of the major problems which continue to plague technical occupations: employment shortages, quality problems, and sagging productivity. To date, however, little is known about the nature of skill in technical work, or where and how technical skills are acquired.

Until recently, employers and policy makers in the technical arena have all too readily assumed that technical skill follows from formal education. As a result, they have approached virtually all labor market problems in technical occupations—skill shortages, low productivity and quality, and the like—as problems of insufficient technical education. It is not, therefore, surprising that they call for educational reform as the primary means of remedying problems related to skill in the technical workforce (Bishop and Carter 1991).

I am indebted to the members of Cornell's Program on Technology and Work: Stephen Barley, Asaf Darr, Bonalyn Nelsen, and Stacia Zabusky, for creating the atmosphere of intellectual exchange that produced the ideas in this paper. I received insightful comments on an earlier draft from Lawrence K. Williams, Jennifer Halpern, and Stephen J. Mezias. Versions of this paper were presented at the 1993 Academy of Management, Atlanta, and the 1994 Conference on Technical Work and the Technical Labor Force, Cornell University.

*What is Skill?*

A classical approach to the study of skill is to treat it as an attribute of individuals. Under this formulation, individuals are said to "be skilled," to "have skill" or to perform in a "skilled" manner. Credentials, including college diplomas, are one indicator of skill and a primary mechanism by which individuals signal to employers their potential to perform competently in a line of work. Credentials provide a means of distinguishing among job candidates in the absence of real knowledge about an individual's abilities on the job.

However, exclusive focus on education as a meter for gauging skill is misleading. Skill is not simply knowledge but also "the ability to do something well" (Attewell 1990; Vallas 1990). Muhammed Ali was a skilled boxer because he possessed an unusual ability to knock his opponents off their feet; Luciano Pavarotti is a skilled opera singer because he hits every note in a score with precision and moves listeners in the process; and Steven Spielberg is a skilled filmmaker because he captures the imaginations of thousands of moviegoers with virtually everything he produces.

An interesting aspect of skill that emerges from this formulation is the concept that individual skill is a highly contextual phenomenon; that is, what constitutes a skilled performance in one context does not necessarily resemble skill in the least in some other contexts. Skill involves a person's ability to respond to the unique demands of specific situations, which is to say that if Ali had boxed Joe Frazier in the same manner he employed on a lesser opponent, he almost certainly would have lost. Similarly, Pavarotti's delivery of a skilled vocal performance depends on his ability to read and react to differences in context—the dynamics of the halls in which he performs, the size of the audience, the particular piece he sings, and the quality of his voice on a given night. Skill almost invariably entails improvisation.

It is partially because skill is so contextualized that we perceive a considerable component of difficulty in any skilled performance. Performances that need not be sensitive to context are said to be easy; people employ more or less the same techniques for walking irrespective of whether they are inside or out, on a paved road or a dirt trail, and so on. But we assume, quite rightly, that contextualized abilities are narrowly distributed throughout the population. Barring physical disability, anyone can walk, but hardly anyone can sing like Pavarotti. Hence, skill has a fair measure of mystery and awe attached to it.

It is precisely for the foregoing reasons that the measurement of skill has proven problematic for researchers. Since it is impossible to "observe" skill directly, researchers tend to rely on proxy measures from which they infer the existence and exercise of skill. This is particularly true in the case of job-related skill. A common metric used in survey research to assess an employee's skill on the job is education, or "years of formal schooling" (Becker 1975; Field 1980). The logic behind utilizing education to measure skill argues that education imparts job-relevant knowledge. The more educated an employee, the more job-relevant knowledge, and hence the more "skill," they are said to possess. This logic forms the basis for hiring decisions in the case of virtually all jobs requiring an education; hence recent college graduates possessing bachelor's degrees are paid more than their counterparts holding only associate's degrees (Berg, 1970). An alternative to the education-centered measurement of skill is the use of output measures such as productivity and quality. The assumption here is that skill translates to performance: the more skilled a person is, the more productive they are and the higher the quality of the results of their labor.

There are both epistemological and pragmatic problems with the use of education as a proxy for skill. On the epistemological front, equating education with skill blurs the meaning of the concept of skill by reducing it to formal knowledge. While such a reduction may be justifiable in the case of skills residing largely in the possession of formal knowledge—of which strikingly few cases come to mind—it excludes from the discourse on skill those occupations which require no formal training or those occupations whose skills cannot be easily inferred from labor process inputs (education) or outputs (productivity, quality).

As a case in point, consider B. B. King, the blues guitarist and singer. King never received formal training on the guitar, and, in fact, claims to be so unschooled that he finds it impossible to play and sing at the same time. Moreover, from a formal technical standpoint, his technique is poor; he employs awkward and even incorrect fingerings on the guitar, lacks the ability to read music, and violates rules of music theory. If one were to employ formal labor process inputs to assess King's facility on the guitar, one would undoubtedly conclude that King is unskilled. Consideration of labor process outputs would lead to a similar conclusion; productivity is an irrelevant measure for assessing a musician's skill (some of history's greatest musicians were notoriously unprolific), and quality is an elusive, largely subjective factor where music is concerned. And yet few would dispute that B. B. King is a hugely skilled musician, worthy of his reputation as "The King of the Blues."

*Skill and Policy.* The obfuscation of the concept of skill through the use of proxy variables has serious policy implications as well. Initiatives aimed at managing problems of quality, productivity and technical manpower are almost always predicated on the measures, but not the meaning, of skill. Too often, employers and policy makers cry for more education and more training without understanding the actual meaning of skill in the context of a given technical occupation. Therefore, they run the risk of simply requiring more of an education that is misaligned with the actual requirements of a job. As Randall Collins (1979) demonstrated, the link between formal education and skill in many occupations is at best tenuous. If there is a disjunction between the meaning and measurement of skill, then initiatives predicated solely on education and training could lead policy makers to implement solutions which, ironically, exacerbate the very problems they seek to eliminate.

*Skill as Practice.* The point of the previous discussion is that there is a disjunction—a conceptual leap of faith—between the meaning and measurement of skill. Education and training do not equal skill; they are merely indicators of skill. Similarly, even productivity and quality indexes may not capture skill. There is a danger that in equating skill with surrogate variables, researchers divorce the concept from work practice—what it is that people actually do on the job—and hence obscure the meaning of the concept. Skill resides not in the inputs or outputs of a labor process but in the process itself. Inevitably, skill resides in a performance.

Therefore, a remedy to these epistemological and pragmatic shortcomings can only be found in the study of work practice—that point in the labor process where skill is actually articulated. Exclusive focus on the metrics of skill in technical occupations have diverted researchers' attention away from the actual practices and processes by which technical work is accomplished. We simply know very little about what technical workers do on a day-to-day basis. The study of work practice offers more than simply a means of understanding what technicians do. A practice-based view of skill serves to bridge the gap between exogenous and endogenous labor process variables by assessing firsthand the articulation between education (what technicians learn in school), skill (the competencies that their work actually requires) and quality (the end products of a technical labor process). As an example of how a practice-oriented approach to skill might shed light on the nature of skill in technical occupations and lead to more effective policy, consider the recent attempts to regulate the quality of U.S. medical labs through minimum personnel standards.

## MEDICAL LABORATORY TECHNICAL WORK

Medical laboratory technical workers, commonly known as "medtechs," are technicians in hospital settings and physicians' office labs (POLs) who perform clinical tests and prepare slides of human body fluids and tissues which are used by pathologists and other physicians to diagnose and treat illnesses. Medtechs employ a variety of tools and techniques—some automated, some manual—to render test results of various kinds from the patient specimens that enter the lab. They take a raw material, such as blood, and utilize microscopes, centrifuges, electronic analyzers, reagents, and a variety of other materials and tools to perform a battery of tests, including counts, identifications, assays, and diagnostics.

There are essentially three types of medtechs. Table 8.1 presents the various educational routes to a job in a hospital laboratory.

Some technicians possess only a high school diploma combined with on-the-job training. A far greater number hold associate's degrees in medical laboratory science, chemistry, or biology; these technicians are called medical laboratory technicians (MLTs). Finally, some medical laboratory workers possess bachelor's degrees in medical laboratory science, chemistry or biology; they are classified as medical technologists (MTs). In addition to their formal education, both technicians and technologists may seek certification from the Board of Registry of the American Society for Clinical Pathology (ASCP). The ASCP registers several classes of lab workers, including MLTs and MTs, cytology technicians, histologic technicians, and phlebotomists (Held, 1991). In order to become Board certified, technicians and technologists must satisfy credit hour and course work requirements and demonstrate their technical competence by passing a national certifying exam.

### *What Do Medtechs Do?*

Medical laboratory work is organized into six areas of specialty: hematology, chemistry, microbiology, cytology, histology, and blood bank work. Technicians in hematology utilize sophisticated electronic analyzers to study blood; the most common test in hematology is the complete blood count (CBC). Hematology typically encompasses a vast array of additional tests, including urinalysis, serology (pregnancy tests, rapid plasma reaction tests, antinuclear antibodies, tests for mononucleosis), and coagulation studies (tests for blood clotting time, tests for clotting factors in the blood). Technicians in the chemistry room also use electronic analyzers, in this case to study the chemical constituents of blood and other body fluids. In addition to the routine chemistry screen, chemistry technicians conduct drug tests, toxicology studies, cancer marker studies, blood gas analysis, thyroid studies, and radioimmunoassays. Microbiology technicians conduct

TABLE 8.1
Types of medical laboratory workers

Non-degreed technicians
- High school education plus on-the-job training
- Military training

Medical laboratory technicians (MLTs)
- Associate's degrees in medical laboratory science, chemistry or biology
- 2 years' coursework

Medical technologists (MTs)
- Bachelor's degrees in medical laboratory science, chemistry or biology
- 3 years' coursework and 1 year practicum

various procedures to isolate and profile various organisms living in body fluids and solids. Cytology technicians make slides of individual cells which are used in obstetric and gynecologic diagnostics; their stock in trade is the pap smear. Histology technicians use manually operated microtomes to cut paper-thin cross sections of human tissue. After cutting, they affix the cross sections to microscope slides, stain them, and deliver them to pathologists, who examine the slides for the presence of disease. Blood bank technicians type and store blood samples.

Although most lab procedures traditionally were done by hand, scientific and technical advances have, over time, occasioned a radical restructuring of the content of many laboratory jobs, particularly those in chemistry and hematology. Specifically, advances in digital technology have allowed computer scientists to program into machines much of the formal knowledge required of chemistry and hematology technicians at the bench. As a result, much of the work which technicians previously performed by hand is now done quickly and accurately by machines. Automation has brought with it fundamental changes in the work performed by most techs on a day-to-day basis. Technicians in highly automated labs are now as much monitors as anything else, responsible for the care and feeding of machines. While histology and cytology technicians have been spared this fate, physicians tend to perceive technicians in chemistry and hematology as semi-skilled, often referring to these techs as "button pushers."

## REGULATING QUALITY IN THE MEDICAL LAB

The 1980s saw a groundswell of concern over what appeared to be diminishing quality in the nation's medical laboratories. Scores of articles in the popular press and on the nightly news painted a fearsome picture. Horror stories of our medical labs—botched test results, improper handlings of specimens, cases of fraud, and AIDS transmission resulting from technician carelessness—led Congress to launch a probe into the problem of laboratory quality. Upon investigation, Congress concluded that, in most cases, human error was the cause of the decline in laboratory quality. Congress's findings implied that poorly educated and poorly trained technicians were turning out inaccurate lab results, and patients were at risk as a result (Rosenberg 1991).

In response to the growing concern, in 1988 Congress introduced the Clinical Laboratory Improvement Amendments, otherwise known as CLIA '88. CLIA sought to improve the state of laboratory quality by regulating the educational requirements of medtechs. In its proposed form, CLIA was to enhance lab quality by bolstering the educational level of the technical labor force.

CLIA charged the Health Care Financing Administration (HCFA) with the arduous task of evaluating the entire scope of existing lab tests and classifying them in terms of their complexity. Complexity was defined as the likelihood of error associated with conducting a test, coupled with the potential harm to patients if errors occurred. As a result of this exhaustive process, CLIA established three levels of test complexity into which all lab tests were classified. The simplest lab tests—a dipstick urinalysis, for example—were classified as waived tests. Only a small fraction of all lab tests fell into the waived category. Level I tests were those evaluated as moderately complex. Approximately 60–70 percent of tests fell into the Level I category. Level II encompassed highly complex lab tests, which accounted for the remaining 30–40 percent of all tests.

CLIA tied lab testing personnel standards to its revamped complexity assessments. Previously, personnel standards were largely the prerogative of individual lab directors; hence, wide variation existed in standards from lab to lab. Standards were particularly low in rural laboratories, where directors often staffed technical positions with high-school educated or military-trained technicians. Under CLIA, all labs would be required to adhere to minimum personnel standards for every level of test complexity. Waived tests, the simplest of tests, required no minimum education. Moderate-complexity Level I tests could only be performed by personnel possessing at least an associate's degree; high-complexity Level II tests required a bachelor's degree. In short, CLIA represented a dramatic upgrade in the educational requirements of laboratory personnel.

Although Congress and the medical laboratories never directly examined the relationship between technicians' education, their skill, and the quality of lab work, CLIA nonetheless contains implicit assumptions about these relationships. The primary assumption that underpins CLIA's attempt to legislate quality through education is that there is a direct and positive relationship between the level of technical education and technicians' skill. If the fundamental assumption underlying the CLIA effort were correct—that technical education is positively related to technicians' skill—then we should observe several characteristics in the social organization of hospital labs: (1) MTs should perform different tasks than MLTs; (2) MTs should be paid more than MLTs; and (3) MTs should exhibit more skill, and hence make fewer errors, than MLTs. It was, in part, to investigate such questions that I undertook an ethnographic study of a medical laboratory. My aim was to articulate the nature of skill in the work of medtechs, to uncover the sources of that skill, and finally, to assess whether an education-centered response to quality problems was justified.

## DATA AND METHOD FOR THE STUDY

To carry out this study, I spent a four-month stint as a participant observer in the medical laboratory of Hilltop Hospital, a medium-sized hospital located in a rural area of a northeastern state. The laboratory primarily served Hilltop's attending physicians but also competed with independent labs for the business of physicians located in the surrounding community. The lab's technicians were assigned to one of six specialized technical areas: (1) hematology, the study of blood; (2) chemistry, the examination of chemical constituents of body fluids; (3) microbiology, the isolation and identification of microorganisms in the body; (4) histology, the preparation of tissues for examination under a microscope; (5) cytology, the study of individual cells; and (6) the blood bank, the storage and typing of blood. These areas were commonly referred to as "rooms" or "sections" by hospital staff.

The lab hierarchy was relatively flat and its accompanying authority divisions were clear. At the bottom of the authority structure, working in support of the techs and the physicians, was a small clerical staff, responsible for the storage and retrieval of laboratory requisitions and reports. Above these clerical workers were the techs themselves. Section supervisors, most of whom were former bench technicians who had been promoted to managerial positions, oversaw the work of the technicians in the various sections. Although the section supervisors were responsible for hiring, firing, scheduling and monitoring the techs, they routinely performed many of the same procedures as their subordinates, a practice that rendered them virtually indistinguishable at first glance. Far greater formal control rested with the laboratory manager, who exercised considerable influence over personnel decisions, pay increases, performance evaluations and general lab policy. Finally, the lab employed two hospital pathologists, both MDs, who assisted attending physicians by interpreting test results and slides. While the formal organizational relationship between the pathologists and the lab manager was unclear, there was little doubt that the pathologists had a significant voice in decisions bearing on all aspects of laboratory procedure, from its dress codes to its technical protocols. Pathologists frequently interacted with technicians in order to exchange information about test orders and results but rarely discussed technical procedures with the techs.

I observed at the lab three days per week for a period of four to six hours each day. Over the course of the study, I observed all shifts in order to capture any distinctions in the lab operation at different times of the day. In addition to observing, I also interviewed several of Hilltop's technicians in both structured and unstructured formats. Often, these informal interviews stemmed from casual conversations with technicians and technologists in hallways, over coffee, or in the course of the social events that techs held, including birthday parties, going-away parties and National Lab Week celebrations. Structured interviews consisted of questions aimed at following up a line of thought, an issue or a concept raised in earlier observations, or as a means of testing tentative research hypotheses derived from field data. While the bulk of the interviews were conducted with technicians, I interviewed both pathologists and the lab manager on several occasions and in several different contexts. Although I typically recorded both observational and interview data as jotted field notes (which I typed and expanded once off site to incorporate data still fresh in my memory), I occasionally taped conversations in order to capture highly technical data that I might otherwise have missed.

The lab was physically divided into several rooms, each housing a technical specialty. Over the course of the study, I rotated from room to room, devoting several weeks of observation to each. This intensive "room-centered" approach enabled me to capture a more detailed understanding of the technical aspects of each type of lab work. Although the technicians rarely left their assigned rooms, there were occasions when their tasks took them outside the lab. Typically, this occurred when technicians were called upon to assist physicians with test procedures conducted in patients' rooms or in special rooms located on patient floors. I frequently trailed technicians on their excursions from the lab. Over the course of the study, I observed (among other things) histology technicians working in the morgue and hematology technicians assisting physicians with bone marrow extractions.

## THE SOURCES OF ERROR IN A HOSPITAL LAB

Medtechs were quick to point out that anyone, MT or MLT, can make a mistake in the lab. Moreover, they argued that the factors which enable a tech to avoid errors have relatively little to do with formal education. Thus, they believed MTs to be no better prepared, from a formal educational standpoint, to avoid errors than MLTs. Technologists, when asked how they utilized their additional two years of formal education on the job, almost unanimously downplayed the significance of the junior and senior year. According to one technologist, "The extra two years don't really matter that much. It gives us a little extra money, and you have to have an MT degree if you want to be a supervisor. But really, it's just a bunch of boring courses that we don't really need to do the job well." How, then, do medtechs manage error in a hospital lab? To approach this question, it is necessary to consider first the sources of error in the lab.

Errors occur when incorrect results escape the lab and make their way onto patients' medical charts. Errors take the form of false positives—test results that incorrectly identify negative results as positive ones; false negatives—positive results incorrectly identified as negatives; and incorrect counts—for example, an inaccurate complete blood count (CBC).

There are three broad sources of error in the medical laboratory labor process. *Pre-analytic errors* are procedural elements that go wrong before a test is conducted, *analytic errors* are elements that go wrong during the actual test, and *post-analytic errors* are elements that go wrong after the test is conducted.

*Pre-analytic Errors*

Several steps exist in the labor process before technicians and technologists actually combine materials, tools and techniques to conduct a test; they comprise what is commonly known as the pre-analytic phase of lab work. At the pre-analytic phase, specimen collection and test requisition take place. Several things can go wrong during these initial steps. First, there can be problems with test requisitions. Sometimes physicians order incorrect tests, inaccurately specify some aspect of a test order, or forget to order a test altogether. In addition, test requisitions sometimes fail to reach the lab, particularly in labs that use paper requisitions, which can be dropped, misplaced, or otherwise lost. Techs refer to problems with test requisitions as "bad orders." According to techs, the physicians and nurses who actually order tests are the culprits behind "bad orders."

"Bad samples," samples that are somehow unfit for testing, make up a second type of pre-analytic error. There are several types of bad samples. Bad "draws" are blood samples whose properties lead to testing inaccuracies. Blood samples that are too old are the most common type of bad draw, but other types of bad draws include blood samples taken from the wrong patient, samples that are too small to get an accurate test read, or samples that are somehow lost en route to the lab.

Most bad samples are the result of phlebotomist error, according to technicians and technologists. They report that it is not uncommon for phlebotomists to enter a hospital room and "draw blood from a roommate" rather than from the patient requiring the blood test. Sometimes, however, patients themselves are responsible for bad samples. The following excerpt from my fieldnotes illustrates the case of a patient who failed to follow pretest instructions from the lab.

> Kay went about the procedure for a sperm count, explaining it to me as she went along. "This guy's sperm count of 62 is low: 80 to 120 million sperm cells per cubic centimeter is normal. The shape of them is pretty normal, though—he's only got about 3 percent curlytails." I asked her what can go wrong with a count like this, and she replied, "The biggest thing is that sometimes we don't get a large enough sample to count 100 cells, which is required. You see, patients aren't supposed to have sex for three days prior to a fertility count. But sometimes we get a really small sample, and it's not too hard to figure out that the guy had sex in the last three days. It's kind of embarrassing, because one of us has to get on the phone and have him come do it all over again. . . ."

*Analytic Errors*

Errors potentially creep into the testing process during the actual conducting of testing procedures. There are two sources of analytic error. The first occurs when machines, which perform the bulk of the testing in hematology and chemistry, malfunction. Traceable machine malfunctions—ones that technicians can diagnose—include tubing problems, internal clogs, and power failures. Non-traceable malfunctions are ones without discernible origins. When techs are not able to trace the cause of machine malfunctions, they typically report that "the machine went nuts," "the machine burped," "the machine is confused," or simply that "the machine is having a bad day."

It is at the final analytic phase of the labor process that technician error is most likely to occur. There are two basic kinds of technician error. The first is a failure on the part of the technician to follow proper testing procedure. Although it is rare, techs do at times omit steps from testing protocols, or simply botch some element of a test procedure. However, techs view procedure as routine and therefore consider it an unlikely source of error. A far more likely source of technician error occurs when techs incorrectly identify or classify cells under a microscope. This error occurs most frequently in hematology, where technicians are required to count and identify various kinds of blood cells. At times, technicians encounter cells that are difficult to identify. In the following excerpt from my fieldnotes, a technician incorrectly classified cells as normal, when in fact they were abnormal and indicative of malignancy:

> Sally and Kay were showing me a slide of a patient's spinal fluid when the pathologist came in the lab. He began to explain the patient's diagnosis to me in terms I didn't understand, except to realize that a malignancy was discovered. This particular smear had attracted a lot of attention because a tech on the "off-hours" shift had incorrectly classified the cells. Lymphoblasts (abnormal cells) had been reported as monos (normal). The pathologist discovered the error upon reviewing the smear that morning. He brought the slide back into the hemo section and left it on the main worktable accompanied by a sheet of paper upon which he had written: "Smear reviewed. Mononuclear cells are large lymphoblasts consistent with malignant lymphoma. Note to techs: these are clearly abnormal cells. To classify as monos implies that they are normal monocytes. Better to report as abnormal cells if you're not sure—suspicious for malignancy pending review, etc."

*Post-Analytic Errors*

Even after technicians conduct lab tests, the potential for error persists. Several sources of error reside in the steps that make up the post-analytic phase of the testing process. During the post-analytic phase, technicians use computers to deliver test results into electronic files which physicians access when diagnosing and treating their patients. There are two major sources of error at the post-analytic phase.

First, any number of reporting errors may occur during the transfer of test results into computer files. Technicians may inadvertently transfer results into the wrong files, or computer errors may send results to the wrong file. Despite technological advances in managing patient files, it is a surprisingly common error.

Hence, there is a considerable movement in lab quality assurance to consistently match test results with patients. As one pathologist said, "We want to make sure that Patient X's sample is the one that ends up in Patient X's file."

In addition, technicians sometimes make typing errors when keying test results into the computer. Computer-generated test results are electronically fed to patients' files, but manual tests, such as urinalysis, are entered into files manually by technicians. To do this, technicians access patients' computer files and key in test results, typically a series of numbers and codes. Technicians report that keyboarding errors are particularly likely when the volume of work is heavy and when they go for lengthy stretches of time without rest breaks.

Even when correct results make their way into the correct files, physicians sometimes err in their reading of test results. At times, physicians simply refuse to accept perfectly accurate test results, typically because the results are unexpected. I encountered several cases of physician "nonbelief" during my stay in the lab. On one occasion, a patient's hematocrit count had increased significantly from one test administration to the next. Changes of this nature are called "delta checks" in lab parlance, and typically require careful attention to ensure that an actual physiological change in the patient has occured. The incident unfolded as follows:

> Joe was rerunning a spec several times and I observed aloud that I hadn't seen so many checks for accuracy before. Joe explained that he was manually rechecking a result he had earlier produced on the machine. Even though the result was correct, a physician had rejected it. "This doctor that I'm dealing with here can get real irate," Joe said. "When this initially happened, the doctor looked at my results and actually called down here and said 'You're full of shit—where's the real result!' He was all over me like a cheap suit, which I thought was real unprofessional. I ended up getting a QA (written reprimand) even though the result was correct! So I want to be sure that I've done everything I can this time to show that the results are right."

Two subsequent test administrations using different methods yielded the same result, confirming the initial, machine-made result. With the aid of a lab pathologist, Joe finally was able to convince the doctor that the elevated hematocrit count was indeed accurate.

## MANAGING ERROR IN THE LAB

Several interesting observations follow from the investigation of error thus far. First, examining error at Hilltop reveals that the source of a large proportion of error resides outside the lab. Despite widespread perception that lab errors originate with lab technicians, the bulk of errors actually originate in the pre- and post-analytic phases of the testing process, both before and after technicians handle specimens. Indeed, several occupational actors may be responsible for "lab" errors, including technicians, physicians, nurses, phlebotomists and couriers working in and around the lab.

Because errors only occasionally stem directly from the work of medtechs, they rarely present themselves in an obvious manner; that is to say, most errors in the lab take the form of test results which, out of context, look perfectly normal. Technicians only discover most errors because they have learned to attend to a variety of contextual cues that indicate that otherwise normal test results are problematic.

Thus, skill in a hospital lab resides in the technicians' ability to detect and control errors that emanate from multiple direct and indirect sources. Moreover, skill is articulated in a fascinating system of complementary practices, methods and norms of behavior that technicians regularly enact in the day-to-day practice of lab work in order to increase the validity of test results and decrease the likelihood that errors will escape undetected. Techs frequently talk about this system of methods and practices when they talk about their work. In the language of the medtech, there are essentially three broad aspects to skill in the lab, each aimed at minimizing error: interpretative skill, troubleshooting machine malfunctions, and improvisation.

### *Three Broad Skills for Controlling Lab Error*

*Interpretative Skill.* The primary aspect of skill in Hilltop's lab was what technicians and technologists called "the interpretative part of what we do." Interpretative skill is the process by which technicians position test results in a context of technical and formal knowledge, tacit and working knowledge, and knowledge of a patient's individual case history (age, gender, specific illness, etc.) to make sense out of anomalous findings. According to the techs, machines serve only to produce results; the medtechs use their interpretative skill to determine whether the results are correct. A technologist described the work of the lab's machines as follows: "The machine just takes the sample and gives you a result—it doesn't know whether it's right or wrong. And that's where we come in. You'll always need a tech in the lab. No matter how much technology advances, they can never take away the interpretative part of what we do."

Examining routine test results for errors is often little more than a matter of verification. As one tech described the process of verification: "Basically the machine gives you some results, and you just have to look them over to make sure they're all right." Frequently, however, the verification process revealed what techs call "ambiguities." Ambiguities took the form of specimens with characteristics that the techs found difficult to identify or test results that were outside of the range of "normal values."

Technicians used a number of interpretative mechanisms in the service of discovering and eliminating anomalous test results. One mechanism the techs frequently employed was collaboration—what they called "working together as a team." Teamwork involved technicians pooling their individual competencies in the task of decoding anomalies. Of the lab's activities, ambiguities were perhaps most common in "differentials," the manual blood counting technique performed by hematology technicians. For manual differentials, technicians used microscopes to identify, classify and count blood cells of different types and in different stages of development. Since the number of potential classifications was vast, technicians, from time to time, had to make judgment calls. When a judgment call was warranted, technicians routinely checked their interpretations by seeking the advice of one or more peers. In this way, technicians pooled their individual competencies to negotiate a collective agreement about the correct way to report ambiguous results.

I observed this process of competency pooling on virtually every day of the study. In the following excerpt from my fieldnotes, technicians collaborated to confirm an exceptional test result and to uncover a potential mistake made the previous day:

> Arlene, one of the techs from the blood bank who happened to be covering in hemo for today, was doing a manual blood count on a specimen. Upon finding a vastly elevated count of a particular kind of abnormal cells known as "atypical lymphs," she began to question the other techs sitting at the table as to what they knew about the patient. (I noted to myself that there is nothing built into the procedure of doing a manual blood count to tell the techs that an abnormal number had been counted—no flag, sign or symbol appears as with the machine.) The subsequent discussion among the techs revealed that this was indeed an exceptionally high count. Karen, an MT, joined the discussion by noting that she had performed a Monospot test on this patient the previous day, and that the result had been negative. In fact, she said, she had run it twice. The other techs were incredulous when they heard this. Apparently, the levels of atypical lymphs they observed in this specimen go hand-in-hand with *positive* Monospot results. As a result of this discussion, the tech who had run the Monospot the day before decided it was in everyone's best interest to run it again. After all, they were on the verge of reporting two inconsistent results, and someone was sure to question it. Subsequent to this, the Monospot was rerun; the result was again negative. Despite the persistence of the inconsistency, the techs reached a consensus to release both the Monospot and the blood count results into the patient's file, having agreed that they had taken reasonable steps to ensure the accuracy of the information.

A second method of interpretation was what techs often refer to as "fitting the pieces of the puzzle together." A lab supervisor described this process (which she valued highly in employees) as "seeing the big picture" when reading test results. She described it as follows:

> What we always do with results, and what I remind other people to do, is to think of the test result as just one small piece of the puzzle. The test result is this little piece, and then you add another—maybe the patient's name and age, and their sex. Then maybe you get another result on that patient and it doesn't check and now you're thinking, "Okay; what's going on here?" So the job is a lot more than just pushing buttons on a computer.

Because errors had serious implications for the techs, solid norms had developed to support and reinforce the practice of interpreting ambiguous cases, as the following interview notes from a conversation with a hematology tech demonstrate. The technicians explicitly invoked these norms when one of their peers violated the practice of interpretation:

> Karen and I got on the subject of error in the lab. She told me that the "off-hours" error described above (misclassified cells), the one which had resulted in a red-inked reprimand from the chief pathologist, had been traced to Emily. She said that whenever such errors occur, techs go "scrambling through the records to find out who ran the test, because you want to make sure it wasn't you." According to Karen, the case at hand involved the failure to identify cells in spinal fluid that were clearly abnormal. In response to this error, Karen had retrieved the original sample and shown it to some techs on the third shift. Eventually it ended up on the main table in hematology, where techs on all shifts had taken turns inspecting it. Apparently, this upset Emily tremendously; she felt that Karen was making her "look bad." But Karen expressed no remorse for her actions, noting that "people get really embarrassed if they don't know something—they don't like to admit that they don't know what something is—but to me, that's nothing to be ashamed of. The mistake is in not having something reviewed, in not asking questions, or [not] trying to find out more if you're not sure what it is."

From this instance, it becomes clear that technicians had adopted an informal practice of public humiliation in order to impart and reinforce the importance of teamwork in negotiating the interpretation of ambiguous test results. Lab supervisors also recognized the importance of interpretative ability and utilized it as a criterion for evaluating technicians. For example, Jack, the chemistry supervisor, when asked about the qualities he seeks in new recruits, said that a tech's formal education and credentials were less important than "an ability to learn and a willingness to say to someone else 'I don't know what something is.'" However, supervisors expressed frustration at their inability to reward technicians directly for such skills, since pay was tied largely to education. As one supervisor explained, "It's frustrating in that right now, we have some wonderful two-year technicians in this department, techs that I would trust any of my results with, but I don't have a way of recognizing them financially, because they don't have the degree."

*Troubleshooting Machine Malfunctions.* A second broad skill for overcoming the sources of error in the lab was the ability to troubleshoot machines. Machine troubleshooting was the process of developing intimate knowledge of machines' inner workings in order to diagnose errors. Technicians who were best at troubleshooting had fashioned a remarkable ability to home in on machine problems by using a combination of sight, sound, and technical know-how.

Although the techs regularly calibrated and maintained the lab's analyzers, the analyzers did malfunction on occasion. Given that the speedy delivery of test results depended on the machines' proper functioning, breakdowns were viewed by the technicians as no small matters. Because physicians require laboratory services around the clock, and because manual procedures existed for the vast majority of lab tests and procedures, technicians' work did not stop for machine breakdowns. However, breakdowns did create unmanageable workloads for the techs not only because manual procedures were far slower than automated ones, but also because malfunctions required that test requisitions and results, which were normally transmitted automatically, had to be posted by hand. In addition, breakdowns elevated the technicians' anxieties because they believed that malfunctions increased the probability of reporting incorrect results. Although procedures for detecting abnormal machine values were in place, incorrect results occasionally managed to slip through the cracks, resulting in either verbal or written reprimands for the technicians.

Because technical malfunctions had such dramatic implications for the work of the techs, skilled status accrued to those techs who possessed the ability to troubleshoot breakdowns. The case of Tom, an MLT in hematology, illustrates this point. Tom was revered by his colleagues for his ability to troubleshoot hematology's analyzers with uncanny speed and accuracy. His troubleshooting skill on the Coulter electronic blood counter, the machine around which the work of hematology technicians focused, was particularly well developed. The following excerpt from the author's fieldnotes demonstrates the nature of his troubleshooting skill:

> When Tom ran the blood sample through the machine, an "ERROR CODE 37" message appeared. He opened the door of the machine and quickly noticed a section of tubing that had become twisted around a component and was preventing the cap piercer from working. Using his thumb and forefinger, he repositioned

> the tubing and closed the door of the machine. He began to explain how he responds to error codes. "I do most of the diagnosing by watching," he said. "I actually open up the machine when a specimen is running and watch what's going on. They're mostly mechanical errors—a pump is clogged or something like that. So you can kind of look in and see what's wrong with this picture. The process of diagnosing errors is a process of elimination. We want to narrow down the range of possibilities to home in on the problem."

Tom had developed this technical acumen largely on his own and without the aid of technical manuals. In fact, the error codes that the machines showed during breakdowns had little meaning to Tom, despite the fact that the codes were explained in technical manuals:

> Tom went on to explain that the Coulter manual has a section detailing the various error codes and what they mean. However, he was unable to find this information in the manual lying atop the machine. He said he rarely uses the manuals anyway, relying instead on "my own material," which turned out to be a worn thick red binder in which he has accumulated hundreds of technical documents, jotted notes and observations pertaining to all aspects of laboratory machine operation. He said it has taken him several years to collect these documents, beginning with his college days and continuing to the present.

Tom valued his red book greatly, keeping it locked away in a private drawer. However, he was quick to share his knowledge with other technicians when breakdowns occurred. Tom's troubleshooting ability had clearly enhanced his standing among hematology's techs, who often called on him when their machines malfunctioned. Recognition of his ability extended even to the pathologists, who had given Tom the unheard-of task of programming the lab's coagulation analyzer within the operating parameters they had established.

*Improvisation and Artistry in Histology.* A final skill, most profoundly evident in the work of the histology technicians and technologists at the hospital, was improvisation—the ability to develop on-the-spot techniques for controlling unexpected variations in test materials. Techs viewed improvisation as a relatively fixed characteristic of individuals in the lab. It appeared to account for a primarily tactile component of skill possessed by only the finest histology techs. As one histology supervisor explained: "There's a certain feel that the best histo-techs have, and really, it's a talent. They just seem to be blessed with the ability to work with a microtome. I've seen perfectly smart, otherwise capable people in a histology room who just never develop that feel no matter how long they've been working. And they probably never will; it seems you either have it or you don't."

On cursory examination, the histology room of the lab seemed almost primitive when compared to the high technology of the chemistry and hematology rooms. Electronic analyzers were almost completely absent from the histology room. Instead, the histology technicians worked with microtomes—instruments designed for cutting extremely thin cross sections of tissue for microscopic examination. Unlike the analyzers in hematology and chemistry, microtomes are operated manually—a deceptively complex human-machine system requiring tremendous hand-eye coordination, patience, and tactile facility on the part of technicians. Because microtomes are extremely difficult to operate, they provided the focus for histology technicians' definitions of skill.

The chief task of a histology technician was making slides from tissue samples—a task known as "fixing slides." Considerable work was involved in cutting whole sections of tissue down to sheer ribbons literally millimicrons thick. The first step in making slides was embedding small bits of a tissue sample into paraffin blocks. The techs in histology viewed "embedding" as the first critical juncture in tissue processing, for it required that they position the tissue in the paraffin so as to maximize the potential for a cut that would yield a representative cross-section of cells. In order to accomplish this, technicians had developed a practice of embedding the tissues "flat side down." However, the techs pointed out that it was often difficult to determine which side of a sample was the "flat side." In addition, most of the techs could think of cases for which they would suspend the rule of "flat side down." Ultimately, the histology techs believed that until they actually cut a sample on a microtome, it was impossible to gauge whether they had embedded it correctly.

Cutting involved a delicate interplay between the microtome's components and the technician's movements aimed at producing a long, wafer-thin cross section of tissue, which the technicians called a "ribbon." The microtome consisted essentially of two major parts: one which dropped a razor-sharp blade through a vertical cutting path (as in a guillotine) and another which pushed the paraffin block toward the moving blade. A technician would control the actions of both the blade and the paraffin block by rotating two handles on the sides of the microtome. When the paraffin block reached the blade, the movement of the blade through the paraffin created an unbroken ribbon of tissue which the technician then guided away from the microtome with the aid of tweezers. Because ribbons were so thin, they were susceptible to static electricity, slight air currents in the room, and even the smallest of unintended movements from the tech, all of which could instantaneously reduce a sample to a useless jumble.

If the ribbon survived, the tech removed it from the microtome and floated it in a small bath of water where it could be more easily captured on a glass slide. Ribbons were frequently lost at this point. A technician might inadvertently drop or twist the ribbon when moving it from the microtome to the water bath (which was positioned directly beside the microtome), or if the tech failed to float the ribbon with a swift, precise motion, it would break apart on the water, and the block would have to be recut.

The technician's final bit of artistry was selecting one of the cross sections of tissue floating on the water for affixation to a slide. Technicians captured their chosen sections by positioning a slide under the section and then quickly lifting the slide to meet the tissue which, upon contact, would adhere to the glass. The potential for tissue distortion at this stage of the process was great. If the technician failed to select a section quickly, the ribbon would dissolve in the water; if the tech did not lift the slide swiftly and precisely, the ribbon would break or be mangled. Histology technicians uniformly agreed that determining which cross sections would make the "best" slides was, at best, difficult and that criteria varied from tech to tech. Technicians' differing conceptions of aesthetics had much to do with the decision. According to Peter, Hilltop's histology supervisor, some techs favored representativeness—a section which would provide the "best picture" of the entire tissue—while others sought the section, whether representative or not, which evinced the least "distortion" of the original tissue.

Because slide-making was not amenable to standardized routines for controlling deviations, histology technicians believed that their ability to produce quality slides stemmed from a happy confluence of favorable and serendipitous circumstances. Yet despite their belief in luck, each technician had developed his or her own methods for controlling errors and improving the chances of making a "good cut." Peter was particularly proud of the technique he had developed for keeping paraffin ribbons intact as they were pulled from the blade of the microtome. He found that by blowing on the ribbon with a constant and very light stream of breath, he was able to minimize the ribbon's tendency to twist or otherwise distort. Each technician frequently improvised techniques on the spot in order to counter the unpredictable behavior of paraffin ribbons. At times, technicians improvised by subtly adjusting the speed of cutting, while at other times, they modified the technique of pulling the ribbons from the instrument or floating them on the water bath.

Given the difficulty of cutting blocks, it is not surprising that histology technicians had developed unique and ritualistic work practices which, in some cases, bordered on superstition. For example, Theresa, an MLT, preferred to set the room's radio to album-oriented rock when she cut, while Peter had little tolerance for anything other than classical music when he used the microtome. Rock and roll distracted him while classical lulled her to sleep; in both cases, the potential for disastrous results was great. Similarly, each technician laid claim to a particular microtome and a particular set of utensils, including tweezers and cleaning brushes, without which they preferred not to cut. The techs refused to share microtomes, utensils or even work space. I encountered this norm rather abruptly when, on the first day of observation in histology, Peter announced, "Whatever you do, don't touch Theresa's microtome . . ." She finished his thought: " . . . Or I'll kick your ass." As if to concretize their feelings of ownership, the technicians had adorned their microtomes with racing stickers, photos, cartoons and various other identifying markers. When the delicate balance of ownership was disturbed, the techs became unsettled, a point illustrated when Peter temporarily misplaced his favorite cleaning brush. Though he had several others on hand, he spent the better part of twenty minutes searching for it before returning to the task of block-cutting.

In the context of histology, skill and artistry became virtually synonymous. The histology technicians treated their microtomes with virtually the same reverence a violinist might afford his prized Stradivarius. Indeed, techs viewed respect for their instruments as an essential part of the constellation of factors constituting "good practice." According to Peter, histology technicians have to cut a finger on a microtome "at least once, in order to respect the microtome," before they can be considered proficient at operating the instrument. Just as violinists develop callouses on their fingertips, skilled histology technicians wear their battle scars proudly. In addition to this deep appreciation for the subtleties of the tools of their trade, the histology technicians seemed to take exceptional pride in the mix of artisanal and technical skill they brought to their work. Histology technicians' attempts to construct their work as a craft became at times almost embarrassingly exuberant, as illustrated by the remarks of an MT in a practitioner's journal: "A histology professional is a special breed—a unique and talented artist who can take scraps of raw tissue and skillfully and efficiently transform them into small, artistic masterpieces in a few hours. Using a combination of science, art and mechanics—a conglomerate of scientific knowledge, creative ability, and sophisticated instruments—they make the invisible become visible" (Knight 1991).

## OBSERVATIONS ON SKILL AND PRACTICE

This study reveals that skill in medical technicians' work clearly entails a substantial component of what sociologists have dubbed "working knowledge" (Harper 1987): job-relevant know-how that accrues from experience and multiple opportunities to interpret test results in light of critical contextual cues. Technicians' working knowledge is largely tacit; they speak of it as an ability to "feel" things, to "see" things, or simply to "just know from experience." Kay, a certified MT, described her working knowledge in this way: "The most difficult thing in this job is interpretation. There are some things you simply wouldn't know until you've seen them many times. Like some cells, for example, are hard to classify, and the only way I know them is because I have a lot of experience looking at slides. I've seen them many, many times."

Technicians emphasize this informal aspect of skill when they talk about their work because it is working knowledge, and not formal knowledge, that enables them to avoid most errors. Techs call on their accumulated experience and contextual knowledge in order to flag potentially problematic test results. Formal knowledge appears to be a figural element of skill in the realm of the routine, but working knowledge enters the fore when techs encounter novel situations on the job.

This view of skill as having an improvisational quality is in stark contrast to lab managers' and administrators' conceptions of technician skill. Recent waves of Total Quality Management in hospital labs have emphasized standard operating procedures and protocols as a means of avoiding errors and producing quality results. As one technician commented, "They've got a protocol for everything we do in here, including how to go to the bathroom." Technicians, however, believe that standard operating procedures do little to ensure quality results, since they do not allow technicians to vary their practice to account for changes in the host of contextual factors that play into the testing procedure. As techs frequently comment, "What works on paper doesn't always work in the real world." Indeed, most techs believe that strict adherence to protocols alone may even increase the likelihood of committing errors. According to one lab supervisor, "There's a lot more to this job than simply following standard operating procedures. It can be, and believe me, some people treat it that way, but if you really just sit back and let things slide, you're bound to make mistakes."

All of this is not to suggest that formal education is irrelevant. There is clearly a base of formal math and science that technicians routinely utilize on a day-to-day basis in the service of carrying out their work. Technicians use math skills regularly when they "titer out" solutions, add reagents to diagnostic tests, or calculate percentages of cells in a blood smear, to cite but a few examples. But virtually all technicians report that on the job, they employ only a small fraction of what they actually learned in the classroom.

The current obsession with technician education forces a myopic view of the problem of lab quality. It inevitably leads to the question: "Who commits the greater number of errors: two-year educated MLTs or four-year educated MTs?" I submit that perhaps this is the wrong question. In fact, examining the work of medical lab workers suggests that the sources of the quality problems that appear to emanate from the lab may in fact exist elsewhere. If one ultimately seeks to improve the quality of results that come from the lab, and not the quality of the lab per se, then one is forced to cast the issue of quality in far broader terms than has been done. If we consider the system of related occupations, departments, and jobs that combine in the production of lab results, we find that quality is a far more complex problem than current thinking allows, and that "blame" for mistakes is in fact distributed throughout the system.

Technicians' identities are firmly rooted in the notion that they take responsibility for managing other groups' errors. Techs thus perceive remarkable value in their work, value that they feel goes unrecognized and unrewarded. There is a widespread feeling among the technician community of being misunderstood, underappreciated, and incorrectly blamed for others' mistakes. These elements of technician identity are captured in this comment by the chief pathologist at Hilltop's lab:

> A lot of our time is spent following up alleged mistakes. We like to uncover the source of errors, even though in a well-run lab, the lab is not responsible for most mistakes. But we're very careful to diagnose pre- and post-analytic errors because the lab always gets blamed for other people's mistakes. In the medical community, it's what's known as "the infamous lab error." Doctors and nurses lay the blame on the lab because of convenience; we work behind the scenes and don't have contact with patients and their families on a regular basis, so it's easy to lay the blame on us.

The critical question in relation to technician skill, then, is not so much *who* commits more errors (there is ample evidence to suggest that physicians and nurses may be as guilty as MLTs and MTs) but *how* technicians manage to avoid errors in the first place. We should take heed when medical laboratory workers with ten, twenty and thirty years of experience tell us that "education is no substitute for experience," and that what they learn in the classroom, while critical, represents but a small fraction of what enables them to perform their jobs well on a daily basis.

My observations suggest that the ability to avoid errors is a skill acquired largely on the job, having relatively little to do with formal education. If this observation is correct, then it is not education but experience that will distinguish technicians' ability to produce quality results. Experience being equal, MTs and MLTs should make roughly the same numbers of errors. Moreover, if working knowledge born out of experience is the critical factor enabling techs to voice errors, then CLIA or any other legislative initiative aimed at bolstering the educational requirements of technicians will do little to improve the quality of results coming from our hospital labs. Simply requiring more or less formal education will not serve to transmit to technicians the aspects of skill that enable them to avoid errors on a regular basis.

These hypotheses are certainly borne out by the statements of the lab technicians and managers I interviewed. Technicians, technologists and lab administrators alike all point to the invaluable role of experience in producing quality results. Indeed, many lab managers report that they prefer to hire experienced

MLTs over techs possessing newly acquired MT degrees. As the lab manager of a large urban hospital laboratory reported:

> We have some incredibly talented people here that are categorized as MLTs, even though they never really finished a two-year program. They got jobs fresh out of high school and they were trained and they've been here for twenty years, and they are great technicians. I trust them with any of my results and I know they would be accurate and precise. And when CLIA first came down the pike, I was afraid they were going to lose their jobs. I have MLTs with fifteen to twenty years of experience that really are much better than MTs who have maybe only two years of experience. Like they say, experience is your greatest teacher and I still think that is very true. As you see things coming up over and over again, they just make more sense to you.

Ideally, then, we must assess the relative contribution of education and experience to technicians' skill—that is, to their ability to manage error. Questions such as these imply counts: do more experienced techs make fewer mistakes than better-educated techs? Clearly, ethnography is not the preferred method for answering questions involving frequencies, since it will rarely produce samples of a size sufficient for statistical analysis. But the ethnographic data from this study clearly implies that traditional ways of thinking about the quality problems in medical labs are shortsighted. Quantitative studies are needed to pursue the lines of inquiry suggested by this research. Whatever the method, the ultimate aim must be determining the nature and amounts of education and experience necessary to improve the quality of results coming from our medical labs, rather than overemphasizing the importance of the classroom at the expense of the laboratory.

# 9

# ENGINEERING EDUCATION AND ENGINEERING PRACTICE: IMPROVING THE FIT

*Louis L. Bucciarelli and Sarah Kuhn*

As the twentieth century nears its close, the world of engineering work is undergoing dramatic change. American technology, preeminent after World War II, is being challenged by foreign competitors. U.S. firms find themselves facing stiffer competition, and many have moved to reorganize and restructure. Reduced military spending challenges firms dependent on government contracts to seek new markets and new ways of doing business. The new technologies that have made global production and global capital markets possible are also transforming the nature and content of engineering work and prompting questions about the prevailing engineering curriculum.

We claim that there exists a serious mismatch between the content, structure, aims, and values of contemporary engineering education and what is required of individuals who wish to function effectively as engineers in today's industrial settings. The view, widely held by the public, that engineering is fundamentally the application of scientific principles is also deeply ingrained in the engineering curriculum. This way of thinking overlooks a significant component of an engineer's actual work, and perpetuates the mismatch between engineering education and what engineers actually do—a mismatch that is growing more pronounced as political and economic changes alter the practice of engineering.

## CHARACTERISTICS OF ENGINEERING PRACTICE

To demonstrate the mismatch, we will begin by describing the essential features of engineering practice: a difficult challenge because the variety of tasks engineers perform is diverse. Even the labels that engineers carry point to significant distinctions—a mechanical engineer is different from an electrical engineer in important ways. Work in a private firm is different from work on a one-of-a-kind civil engineering project. Engineers design the heavy machinery of production, redesign the simplest of household appliances, work to get people to the moon, write software to guide missiles to their targets, maintain instrumentation, diagnose failures, and schedule the dispatch of transit vehicles, power, and raw materials to be processed. Some work in tight hierarchical organizations; for others the formal organization is transparent. Many take on managerial functions with time.

With this in mind, we first describe what we see as the common features of engineering that reach across all modes and specialties. We distinguish two quite different worlds which an engineer must inhabit: what we call the object world on the one hand, and a larger, process-oriented social world on the other. Each has a different character, and the ability to operate effectively in both worlds is essential to effectiveness as a practicing engineer.

### *Work within "Object Worlds"*

We use the term "object world" to refer to the domain of thought and action within which participants in engineering design move and live when working on any specific aspect or subsystem of the design.[1] Work in the object world consists of intensive interaction with the object of the design, with its physical properties, and with how its properties and behaviors might be embodied in order to accomplish the design objectives. Within object worlds, participants in design—mechanical engineers, technicians, marketing managers, project leaders—all go about making up scenarios about things and principles, physical concepts and variables and how they relate. For example, an engineer responsible for sizing the structural members of an automobile chassis will describe the "behavior" of the object in terms of its stiffness in torsion and bending, will express concern about the levels of stress it can "tolerate" before failure might ensue, and will construct an analysis to predict the structure's resonant frequencies and modes of vibration when fully loaded and traversing a bumpy road. (The bumpy road must also be realistically constructed.) The engineer's scenario about the "performance" of the frame can take a dramatic form depending, in part, upon the au-

1. Portions of this section are drawn from Louis L. Bucciarelli, *Designing Engineers*, (Cambridge: MIT Press, 1994).

By design we mean design in the broad sense. We include participation in the orchestration of plans, procedures, and instrumental details of transforming ideas into working prototypes, then into products. But design also includes negotiation of performance specifications with customers, and of constraints with suppliers; it includes trusting, and explaining to others in ambiguous situations. Design is all of this and more.

dience. The object seems to come alive in these descriptions, to develop a character all its own. These scenarios are continually modified and elaborated upon, sometimes made concrete in actual bits and pieces of prototypical hardware, sometimes recoded as mathematical models to be experimented with on the computer. Within object worlds the engineer tests, modifies, and strengthens the scenario and, in the process, assimilates and appropriates the object.

Although work in the object world requires creativity, the goal of storytelling and scenario making is to achieve closure: to arrive at a design that is fixed, repeatable, stable, unambiguous, and internally consistent. Object world thinking is thinking about the rigidly deterministic. It is here, working within object worlds, that science-based engineering carries full weight. Object world thought is also abstract and reductionist. Models of the sort prevalent in science, and in large part derived from science, are essential to work within object worlds. The engineer's ability to abstract from a concrete situation, to see an object as a collection of forces, or as a network of ideal current generators connected in series and in parallel, is key to problem solving and to managing complexity within object worlds. One of the crucial skills conveyed as part of disciplinary training is the ability to look at a design, or at a collection of objects, and to see them as an abstraction to which scientific principles can be applied.

Object world work is, perhaps above all else, work within a discipline. Disciplinary tools are brought to bear, and the designer's thought process is likely to be inaccessible to people from other disciplines. While there is diversity even among members of the same discipline, they will share a common language and a common set of tools—that is, a way of seeing—which allows them to work together on a shared problem at a level of detail available only to those who are members of the discipline.

Object worlds also have conceptual hierarchy. To engineers, nature appears to have structure and to be hierarchically organized, reflecting different levels of scientific abstraction. The general relations that hold among the forces acting on a continuum in static equilibrium are more "fundamental," and hence higher up the hierarchy, than the equations derived from them that determine the stresses in a cantilever beam. The solid state physics theory that describes how the radiation from the sun stimulates the movement of electrons within a thin, doped layer of silicon is conceptually higher up than the model of the behavior of the photovoltaic cell found in a text on circuit theory. Designers must choose the correct level on which to operate at each stage of the design process.

Work in the object world is usually solitary. Although disciplinary colleagues can and do collaborate, much activity is conducted by the engineer working alone. The engineer formulates an explanation, a scenario; he or she then tests it through "interrogation" of the object: conjecturing a cause, making a change, and watching the response of the object. Such activity can be conducted by two or more people together, but solitary work is the norm rather than the exception.

Much of the object world is shared or shareable, but aspects of it are personal as well. The character of the object world is not completely determined by its inhabitant's disciplinary training, although such training does more than any other single feature to define the object world. An individual's work experience, personal history, personality, beliefs and culture all make any given object world characteristic of that individual. This accounts for why even engineers trained in the same discipline will find different solutions to design problems, and why communication—even between disciplinary peers working in the same object world—is often imperfect.

Finally, work in the object world, despite its name, does not always involve interaction with physical objects. The engineer may be working instead with ideas, equations, or mental models. Thought experiments may play as important a role as hands on work with models, prototypes, and materials. The character of the object world is determined by the nature of the work itself, the nature of the thought process and the qualities of interactions, not by the presence or absence of physical objects.

### *Design Is a Social Process*

Work within object worlds is but one aspect of the design process. For design to proceed, engineers must exchange and negotiate with others in the firm, since a complete design is rarely the province of a single individual. Here, different object worlds intersect, generating a sort of work whose character is fundamentally social and process oriented. There is no overarching instrumental strategy for reconciling and synthesizing diverse design interests. To put the claim in the strongest terms, we say that design is a social process. This claim goes against the conventional wisdom that engineering is fundamentally scientific. While essential to design, science is not determinate of the design or of the design process. Of course scientific laws constrain an engineer's options, but constraints imposed by nature are only some among the many constraints binding the engineering design process. Other kinds of technical constraints are more clearly human constructs: the interface conditions among different subsystems, worked out by the project leader in consultation with the chief people on his or her team, are an example.

The performance requirements set by a customer are also constructs subject to change. It is not difficult to lay out performance specifications at the beginning of the design process; indeed, it is standard practice. What is difficult—probably impossible—is retaining those specifications without an ongoing process of modification, clarification, negotiation and joint meaning-making. Specifications that seem clear at the outset are stretched and challenged by the design process itself; ambiguities, incompletenesses, confusions, and contradictions are uncovered as part of the process of discovery that is design.

Still other constraints, unlike those based on scientific law, are normative, and therefore negotiable. Such constraints include regulations promulgated by a government agency, or codes published by the American Society for Testing Materials or the Institute for Electrical and Electronic Engineering. As in any other human construct, there is always a need for interpretation in their application, always more than one way to meet their intent. Constraints are heterogeneous—electrons, cost, regulations, time, imagination. Some of these are more flexible than others; most, if not all, are subject to multiple interpretations.

So too is the division of labor in design ambiguous. Where you place the boundaries between different design tasks, how you break up the entire job of design, is no straightforward process. The boundaries of subsystems—and thus, to an extent, of object worlds—are themselves ambiguous and subject to negotiation and renegotiation. Nor are boundaries absolute even when this negotiation has taken place, since coordination and overlap are both necessary and unavoidable.

Design is thus best seen as a process of communication, negotiation, and consensus-building. No one person, no one object world, dictates the form of the design or even knows the design in its totality.

## CHARACTERISTICS OF ENGINEERING EDUCATION

Like engineering itself, engineering education is a complex and diverse domain. To demonstrate the mismatch between engineering practice and the structure and content of an engineer's training, we will focus on engineering education's obsession with science.

### *The Ideology of Engineering as Science Centered*

The ethos of undergraduate engineering education is science. In making this claim we distinguish between science as practiced in the laboratory and science as taught at the undergraduate level. While the first may in fact engage practitioners in creative and open-ended ways, the undergraduate student's experience in science is more mundane and constrained.

So too for the engineering undergraduate. Indeed, aspiring engineers' first year or so at university is mainly spent taking mathematics and science courses. These are typically taught in authoritarian ways, with the apparent intent of getting students to think, speak, and act in terms of the theories and methods dispensed. The authority and hierarchy of knowledge presented are rarely open to question. Critical reflection on the historical settings of theory development or the modern contexts of its use is, if not openly discouraged by faculty, ultimately devalued by the student as a waste of time and energy. In this respect, the content of the students' course work is very narrowly defined; but in another respect, the content can make claim to universal importance.

As students move into their engineering studies in their second and third years, their learning experiences hardly differ. The emphasis remains on abstract theory and on the utility of mathematics in developing theory's implications for particular problems and phenomena, except now these have a more "engineering" cast. The student continues to spend most of his or her waking hours struggling to find the answers to the weekly assigned problems. There are some differences: the organization of knowledge, still instrumental and reductionist in character, takes a different form. Mathematics and science are used to explain structures, electrical circuits, fluids, computation, and materials. Students now study the workings of cantilever beams, internal combustion engines, photovoltaic cells, and aircraft control systems—in departments labeled mechanical, electrical, civil, and the like.

The curriculum's stress on instrumental analysis is reflected in a textbook for second year engineering students, which states that "the main objective of a basic mechanics course should be to develop in the engineering student the ability to analyze a given problem in a simple and logical manner and to apply to its solution a few fundamental and well-understood principles." (Beer and Johnston 1979: xiii) This engineering educator, and presumably all those who use this textbook, see the primary task as developing the student's ability to think logically, i.e. mathematically, in pursuit of the solution to a given problem. The significance of the phrase "given problem" is that the assigned exercise is not just well posed and soluable within the confines of the abstract setting of the text, but generally has but a single correct response, often listed in the back of the book. The student has no opportunity or responsibility to formulate the problem, muse about context, or explore engineering implications. Their analyses of and solutions to the given problem must be simple and straightforward, utilizing but a few fundamental and well-understood principles.

Constructing problems that allow this kind of response, and only this kind of response, requires a heavy dose of abstraction. The simple stick figures and line drawings displayed in the textbooks represent reality from a very great distance. These figures and drawings are best understood, not as representations of real engineering hardware or systems, but as the thinnest of masks concealing scientific principles which the student must uncover in order to reach the answer. The abstract theory itself remains primary.

Each course is defined by its instrumental content. A course in fluid mechanics follows a route defined by the concepts of velocity and pressure, viscosity and continuity, and energy and momentum transfer. Axioms come first, then the fundamental principles, then an ever evolving sequence of "applications" chosen to illustrate the theories. This knowledge, so essential to the engineering of water supply systems, the design of aircraft, or the layout of chemical processing plants, is wrenched out of context. The hierarchy and authority of theory prevail. Here is precision, confidence and certainty. Abstract theory and the uniqueness of its implications stand alone, disconnected, unsullied by the realities of professional engineering practice. The experience conveys the message that engineering learning is the learning of objective truth, and that the most valued engineering talent is analytical prowess.

Successful students quickly learn how to work within this narrow mode. They learn not to ask the wrong kinds of questions. They accept the process of abstraction without question. Indeed, if a student takes an interest in the broader context suggested by the assigned problem, he or she is most likely to be penalized. In this sense, engineering education teaches students "not to see."[2]

Course work of this sort provides the student with little, if any, opportunity to develop his or her creative talents independently or to gain the self-confidence prerequisite to leadership. It does prepare students to work skillfully in the object world, but it can prevent them from developing capabilities needed for the process world. As important as scientific knowledge is to becoming an engineer, the problem assigning/problem solving experience is one that is authoritarian, single-minded and constraining—and potentially debilitating.[3] It is also an approach that places great emphasis on individual accomplishment *qua* individual; the whole grading system is based on individual achievement and prowess relative to that of peers.

Clear evidence of the dangers inherent in an overemphasis on individual achievement is reflected in the following quote taken from a summary of a 1992 meeting of MIT's Industrial Liaison Program Corporate Advisory Panel: "MIT Graduate: Not as perfect for industry as before. Typical product of MIT education—an excellent individual performer but often considers it just about unethical to use results of other people's work. This attitude must change."

Individual competence remains essential to professional practice. To the extent that this eclipses the individual's capacity to work with others inhabiting other object worlds, however, it renders the engineer dysfunctional.

Dean William B. Streett of Cornell University attributes this premium on individual performance to the military precedents in engineering education. He too claims that this ethos must be tempered if we are to capitalize on the human potential for technical accomplishment in our pre-college youth:

> . . . On the positive side, military traditions of discipline, hard work, competition, and individual responsibility . . . have encouraged self-confidence and in-

2. For a more complete discussion see Bucciarelli 1994.

3. For a discussion of the effects of this form of learning and the conceptual barriers it raises see Kuhn and Richardson 1993. Students report their difficulty and anxiety as they attempt to reconcile their engineering training with a human-centered, skill-based approach to manufacturing.

> dependence, resulting in many generations of graduate engineers who have made important contributions to American life. On the negative side, the rigidity and exclusivity of the military model, typified by the boot-camp mentality of the academic routine, the exclusion of women and minorities, and the almost ruthless weeding out of all who do not measure up, have denied many capable and qualified students the opportunity to earn an engineering degree. (Streett 1993, 9)

The almost total focus on individual competence and achievement in mastering the fundamental concepts, principles, and preferred methods of analysis goes hand-in-hand with the organization of knowledge into specialized, independent curricular compartments. For example, hierarchy of knowledge is extolled in the organization of curricula as well as within the individual course. Mathematics and physics are prerequisite to engineering science; engineering science is prerequisite to advanced study within specialized, professional options; the final year is the time and place for synthesis in a capstone design course. Curriculum discussions dwell mostly on content, asking which chunks are essential, which ought to be available as electives, what subsets of courses are currently relevant, how many of these ought to be made available or required of students. There is rarely any serious attention given to the nature of the student experience, to the contexts of learning and practice.

Of course there is more to engineering students' experience than basic and engineering science courses. Students do laboratory work (though this too might have a narrow cast), probably do some design work, and certainly must study the humanities and/or social sciences. Furthermore, much of student learning and growth might occur on a summer job, while working in an undergraduate research position, or while engaged in other less formal student activities. Still, the prevailing ethos of the student's experience is science.

It has not always been so. The dominance of science in the U.S. engineering curriculum is largely an artifact of the postwar period. Before World War II, engineering education had a decidedly practical bent. Despite the influence of émigré engineers trained in the more theoretically oriented schools of France, Germany, and Russia, and despite the demands of increasingly sophisticated technology, universities' heavy dependence on industry for research money and consulting work kept faculty focused on application. After the war, according to historian Bruce Seely (1993), two additional influences dramatically changed the character of academic research and ultimately the engineering curriculum as well.

First, engineers had been greatly disappointed by their limited role in advanced engineering projects during the war. Scientists, with their theoretical training and research experience, had instead been the key players in wartime engineering efforts. Second, a surge in federal funding for research in engineering science lessened engineering schools' dependence on industry for research support and substantially redirected the focus of the academy. The influence of federal dollars was felt first in research laboratories, but the engineering curriculum soon came to reflect the new focus as well. The postwar period, Seely argues, saw the emergence of two distinct engineering subcultures: most engineers employed in industry continued to be significantly affected by the demands of application and of practicality, while academic engineers concentrated on advancing scientific knowledge.

The distinction prevails today. While the subject matter of engineering research can usually be distinguished from that of science, faculty in the two areas, if seen at work, would be barely distinguishable to an observer ignorant of technical content. Both engineering faculty and science faculty are employed as teachers as well as researchers at institutions of higher learning. Both have graduate research assistants working for them. Both spend considerable time searching and appealing for funds. Both write and publish papers in specialized journals and present the results of their research at conferences. Both review the publications and proposals of their peers, compete for tenure, see their work as ultimately beneficial to all humankind, and sometimes engage in controversies over developments in their respective fields. Both speak of their purpose in much the same way—as bringing ". . . up-to-date knowledge to their teaching, while students who participate in leading-edge investigations develop enhanced research skills . . ." (Grayson 1993)

Like scientists, the primary audience for the engineering faculty researcher's products is other engineering faculty in the same specialization. Faculty come together at the department level to define the undergraduate curriculum. Because of the disaggregation of faculty interests within subfields, individual students shuttle from one course to another through a prescribed set of requirements with little sense of the connections within the discipline and hardly a glance at the relevance of other fields. The prevailing view of faculty in charge is that engineering knowledge is best learned in specialized chunks, one added to another, until the student is filled up.

What we report is, we believe, the essential character of undergraduate engineering education. Of course the student does more—he or she takes humanities courses, probably works part time, perhaps on a task with engineering content, has a summer job in industry or does co-op, and learns about negotiation processes through his or her living group or extracurricular activities. Students also learn about making decisions under conditions of uncertainty and limited resources when deciding which humanities reading to put aside at two A.M. when there is a quiz tomorrow in Fluids. Despite all these opportunities for learning in another mode, the prevailing ideology that defines the official curriculum and guides its faculty is fundamentally instrumental.

## IMPROVING ENGINEERING EDUCATION

The engineering curriculum we have described is well tuned for object world activity, we claim, but this, while still necessary, is no longer sufficient for effective practice within a changing environment where the social process of design is ever more critical. We have moved away from the era when federal support for research sustained the engineering subculture in universities. Nor does government spending on the arms and space races any longer define, for most, the character of engineering practice, the objects and systems engineers produce, or their modes of organization and thought. Today, as military budgets shrink and as product cycles shorten, as competition heats up and quality and customer satisfaction come to the fore, and as teamwork and interdisciplinary collaboration are emphasized, the engineering workplace displays a different dynamic.

If the neophyte engineer is to be prepared for this new world, curriculum renovation is needed. Change, however, is not simply a question of an "imbalance" to be fixed by subtracting an engineering science course and adding a management, design, or writing course. What is called for is an exploration of how even the teaching of engineering science subjects might be transformed and better related to modes of engineering thought in practice.

The distinction we draw between the two domains of engineering practice—the object world on the one hand and the social, process-oriented, context-laden world on the other—is, we believe, helpful in understanding the current shortcomings of the engineering curriculum. Engineering education now excels at preparing students as specialized individuals for work in a particular slot within an object world set in a well-defined organizational hierarchy. Students possess quite sophisticated knowledge of instrumental methods and scientific principles when they leave school, but are severely constrained in their ability to apply their knowledge because they have not experimented with it or tested its limits.

We are by no means the only critics of engineering education. Engineering educators, allied with their industrial partners, and acting with the authority of one or another professional society, have periodically evaluated engineering education and recommended change (see, for example, Grinter, 1955). The shape of engineering education over the past thirty years has been defined, as we mention above, within the context of a dramatic growth of investments in corporate R&D, the Cold War financing of the military-industrial complex, and the opening up of space as a frontier for technological development. The fundamental importance of science and mathematics to developments in these domains was recognized and the designers of engineering curricula responded, discarding applied laboratory and training experiences, fashioning textbooks that stressed the universality and power of theory, and muting the distinction between the scientific research of the graduate schools and the thrust of the undergraduate engineering classroom.

In recent years the curriculum has once again become the focus of criticism. A recent survey, of employers of graduates with bachelor's degrees in engineering found that of eight employee attributes, the one which the greatest number of employers valued highly was teamwork—and engineering graduates were less well prepared in this area than in math and science. Furthermore, more than three-fifths of respondents said that in a revised engineering curriculum it would be most important to give more time to the development of students' communication skills (National Society of Professional Engineers 1992). From the National Science Foundation to *Business Week* to the Accreditation Board for Engineering and Technology, numerous voices have been raised urging a change in how engineers are educated. Not all critics agree on the nature or scope of the needed change, but all agree that the present system is less than optimal.

Prescriptions for change take a variety of forms. Some propose a return to a more "hands on" educational experience; others advocate the systematic and pervasive use of modern computer and information processing technology; still others propose lengthening the time undergraduates spend in school and increasing the number of credit hours they must accumulate before earning an engineering degree, in order to include a course in management or communication and make up for the perceived deficiencies. All of these approaches have been supported to some degree through funding at the national level. Still, as worthy of trial as these suggestions are, we hold that a more fundamental change in both the form and the content of undergraduate engineering education is required in order to prepare students better for contemporary engineering practice.

Because engineering is a social process, value laden and not determined by technical considerations alone, engineering education must move away from its almost total focus on individual intellectual mastery of disjointed, narrow content. It must prepare students to flourish in an indeterminate, ambiguous environment in which object world skills are necessary but insufficient. The shift in corporate settings toward decentralized, highly competitive, resource limited engineering practice makes this change in engineering education all the more imperative. Graduates must still be competent, even excellent, at work in the object world, but they must also be able to convey their ideas to others, work as a team, compete for scarce corporate resources, and see the big picture as well as their own place in it.

An engineering education must prepare students for the complexities and uncertainties of engineering practice, break with the protective shielding afforded by the single-answer problem, challenge them to critically reflect upon what they are doing, and require them to articulate and defend their choices of method and designs in front of their peers and faculty. It should, in sum, recognize and teach students how to deal with context. Students need not become experts in organizational behavior, the negotiation of different interests, or the value systems, abilities and aspirations of the consumers and users of their designs, but they must come to see the legitimacy of the kinds of questions and problems that arise in these areas and their central, not peripheral, place in engineering practice.

These changes are fundamental, and they will require significant alteration to the culture, preparation, and practices of engineering faculty. For some, the transition will be painful, disorienting, anxiety-provoking; yet we believe these changes to be necessary if engineering education is truly to prepare students for all aspects of engineering work, particularly in corporate settings. The precise nature of curricular changes must be decided by engineering faculties themselves, based on their own assessment of their capabilities and interests, and of the missions of their colleges and universities. We cannot prescribe a set of changes that will be right for all schools in all cases. Still, we suggest that there are common elements that all engineering programs should share if they accept our analysis of engineering work. To stop short of these changes, we believe, is to fail to recognize the full depth of the challenge posed by the current gulf between engineering education and engineering practice.

Engineering education should "teach design." By design we mean design broadly conceived. It must include attention to the process of negotiation among the interested parties to a project, as well as consideration of manufacturing, construction, assembly, prototyping, and certainly cost. Codes and regulations are important; safety and environmental concerns are more than legitimate subjects. The underlying forms of explanation of performance remain critical but the focus is on the product and function, not theory or analysis.

The most fundamental characteristic of a design-based engineering program should be the open-ended nature of the student experience. In an open-ended exercise, the boundaries of the problem are relatively free. This means that the student can raise, without fear of scorn or derision, questions that may, for their answer, require the knowledge of disciplines other than the one represented by the faculty member in charge of instruction. In other words, the educational experience is not framed by specialized discipline boundaries, but rather by the task itself. This immediately calls into question the organization of engineering education by specialty. Any reform of engineering education should also acknowledge that in the educational process itself, different constituencies participate with diverse, and not necessarily harmonious, interests. By recognizing the diverse interests of faculty, graduate students, undergraduates and administration, faculty can use the university itself to demonstrate to students the organizational issues present in institutional settings.

Any attempt to redefine undergraduate engineering education must contend with the claim that the engineering curriculum is already so packed with subject matter that adding extra topics would only be made possible by extending the period of study. We believe that the number of engineering subfields covered in a curriculum is not the issue. To create an engineering curriculum that gives adequate weight to subject matter which equips students not only for work in the object world, but also for work in the broader process-oriented world of integration and negotiation, we believe it neither necessary nor possible simply to add more courses to the existing structure. Engineering as a social process must be integrated into the science-based curriculum, not tacked onto it. The changes we believe necessary are deep and fundamental: they are changes in culture, in ways of seeing, in ways of relating to the task of engineering and to its scientific content. The social perspective must be exemplified in the sorts of exercises which students perform, in the kinds of problems they are asked to solve, and in the scope of the courses they are required to pass. An appropriately broad education includes such elements as learning ways to formulate a question, learning methods of estimation and parametric variation, learning alternative ways to verify a result, learning when an analysis doesn't apply or only applies approximately, learning the way theory relates to practice, learning how design relates to manufacturing, and so on. If this material is simply added on to the existing curriculum, it will continue to stand apart in students' minds and will conflict with the message which the rest of the curriculum will continue to convey: namely, that engineering knowledge and action can be compartmentalized and organized according to scientific discipline.[4]

It is not our intent to prescribe or propose detailed curricula that might improve the fit between engineering education and engineering practice. Instead we will look at a typical instance in the undergraduate experience of both student and faculty and explore how it might be recast to meet our objective: the transformation of a culture. In this we will keep the "fundamental" content of the exercise constant, at least from the traditional perspective, but embellish and enrich it so that, from our new perspective, the real content is broadened to include the world of process while making the student experience more active and engaged.

The situation is a problem the student might encounter in a first course in engineering mechanics called "Statics."[5] Typically, problems of this sort ask the student to determine the magnitude of some applied or reaction force acting upon or within a machine, a structure, or even the human skeleton. The applicable fundamental principles are those that ensure the static equilibrium of the machine or structure, namely: (1) the sum of all the forces acting upon the system must be

4. We believe that the same problem exists with science, technology, and society (STS) and engineering ethics courses. As long as they are add-ons to a curriculum pervaded by the absence of human agency, the pretense to objective truth, and the omission of process and organizational concerns, the content of these courses will be at odds with the message conveyed by the rest of the curriculum. For a discussion of how students experience conflicts of this sort, and how difficult they are for them to resolve, see Kuhn and Richardson 1993.

5. Such a course might use the textbook quoted earlier in this chapter.

zero (null resultant force), and (2) the sum of the torques acting upon the system must be zero (null resultant moment).

These principles, or laws of equilibrium, are to be applied to an "isolated system." Herein lies the challenge; constructing a correct representation of an isolated machine or structure, complete with symbolic representation of both known and unknown force quantities, is no simple, straightforward task. There are no recipes for doing this which apply across the board. Yet as puzzling as the traditional problem may be, there remains but a single correct answer, often found in the back of the book.

Let us turn to our specific example. A person is pushing a heavy roller over a level lawn and encounters a small step. The yoke and handle of the roller are oriented thirty degrees to the horizontal. The height of the bump is given, as are the weight and radius of the roller. The problem is to determine the force with which one must push in order to just move the roller up over the step.

The figure in the textbook shows a side view of the roller as a circle positioned on level ground just as it encounters the step. A smaller circle at the center of the roller indicates the axis of rotation, to which the yoke is connected. The yoke and handle appear from the side as a single bar, inclined at thirty degrees to the horizontal. The force the person applies is to be understood as running down the handle.

Note that this is a very abstract representation of a roller. It includes only the minimum information needed for the student to solve the problem. The bump is shown as a sharp step, the small circle at the center of the roller is meant to indicate a "frictionless" pin, and three dimensional effects are to be ignored. (The roller could be a mile wide.) Constructing these problems is no light task; the challenge to the author is to write and draw without including any feature of a "real world" lawn rolling that might trigger irrelevant conjecture and questions and lay open the possibility of responses other than the single answer sought; yet enough must be concealed that the task of isolating and applying the fundamental principles remains nontrivial.

Doing the problem requires further abstraction on the part of the student. He or she must formulate another representation, still more abstract than that of the author, which must clearly identify the relevant known and unknown forces as well as pertinent dimensions. The process, from beginning to end, is reductionist. In addition to constructing a sketch showing the forces acting upon the roller, the student must do some mathematical manipulations to relate the step height to other parameters. If the height, radius and weight are given as numbers, then a single numerical answer is required. If they are left as symbols, then a symbolic representation of the pushing force must be expressed in those terms. Either way, there is only one answer, which must be produced using the aforementioned laws of static equilibrium.

The context for doing the exercise is the following: typically the student encounters the exercise on a weekly problem set, along with several others of the kind. Students are expected to work alone, although some faculty might encourage them to discuss problems among themselves in order to get started. Outright copying is condemned. The model of workmanship is individual; it is unethical to use the work of others.

While the emphasis appears to be on process—modeling the isolated roller, applying the principles of static equilibrium, carrying through the algebra—there is a powerful incentive to get the right answer. If the answer is correct, graders generally give but a cursory review of the student's process. On the other hand, if the answer is incorrect, or if the student submits an incomplete analysis, the grader must spend significant time and energy diagnosing where the student went astray if his or her evaluation is to be more than a cryptic indication of fault (an "x" or negative scoring). While this fuller evaluation might be the occasion for real learning, on the part of the grader as well as the student, the pressures of ongoing course work reduce the likelihood of this kind of exchange, valuable though it might be. Furthermore, within the traditional course structure, the term's problem sets are normally worth, in total, only 15 percent of the final grade. As this particular problem is but one of perhaps forty assigned in the course throughout the term, there is hardly any incentive for the student to look at the grader's evaluation and rework the problem, even if some helpful clue is given. Rewriting, in this sense, is not rewarded in engineering education.

The solutions sheets distributed to students, with their cryptic sketches of critical steps in the solution process, reinforce the message: provide minimal text in producing the right answer. Do not try to explain where you have encountered difficulty; do not indicate what you think might be a questionable assumption. Write in the economical style of the faculty's lecture presentation. Of course, you need not indicate the sources you drew upon in working up your analysis.

Recasting this problem along the lines we recommend, we tell the student: "Your task is to do an analysis in support of the first-cut design of a new, lightweight, mobile hospital bed. You know that the bed will be used to transport patients indoors on caster type wheels over relatively smooth terrain, but there will be some small obstacles and bumps it must traverse without discomfort to the patient. A single attendant should be able to push the bed to its destination. Develop a rationale for fixing the size of the wheels and use it to determine a range of possible diameters."

How does this problem differ from the previous, traditional exercise? How is it the same? We respond, considering three different contexts: the abstract context of the object world of the textbook; the context of the real world of hospital beds, heavy patients, tired attendants, caster wheels, and bumps; and finally, the context of the classroom and the performance of the exercise.

With respect to the abstract world of the text, the two tasks are very similar. The student must isolate the bed (with patient) and show the force exerted by the attendant pushing the bed as the front wheels just start up and over the small bump or step encountered. The isolation of the bed is more complex than that of the roller, and there are a few additional parameters that appear in the equations of equilibrium, but from a traditional perspective and within this context, the two problems are very much one and the same. In posing and working out the roller problem, the real world is to be avoided altogether lest it confuse the student and detract from the abstraction process; here, in sizing the wheels of the hospital bed, the real world must be explicitly attended to. Students will ask about bumps—their sources, frequency, and height. Are they sharp-cornered or rounded and worn? What difference would that make? What is a conservative assumption? What are the normal dimensions of a hospital bed? How much do patients weigh? Where is their center of gravity? With what force might an attendant be able to push?

While most of these queries are appropriate in the traditional sense (that is, responding to them is prerequisite to doing the problem in the abstract world context) they would hardly get the full discussion merited here. Still other questions may spring from a concern for safety or cost—legitimate considerations now, but not before. Some students' questions may appear irrelevant, such as those regarding the color of the floor. Even here, however, the instructor must listen for the possibility of a connection to a legitimate concern in design.

In attending to the "big world," then, the instructor and students must judge and shape the task at hand. Explicit attention must be given to the kinds of assumptions made, to the uncertainty in estimates of important parameters, and to the arbitrary nature of some instructor-imposed constraints (e.g., we will *not* allow the attendant to come around to the front of the bed to lift the front wheels up over the bump).

Ideally, this last dictate will engender heated discussion:

Student: Why can't we allow the attendant to do this?
Faculty: Our design is for normal, frequently occurring step heights.
Student: But how do we know our bed won't encounter such a big bump?
Faculty: We don't, but if we design conservatively, with the attendant pushing . . .
Student: But what if the diameter or cost of the wheel grows excessive . . . ?
Faculty: A good point, but still that's going to be a rare event . . .
Student: Infrequent perhaps, but probable . . .
Faculty: Okay, for the rare event of a bump outside the design envelope, we will allow the attendant to go around to the head of the bed and lift . . .

This dialogue suggests that explicit attention to the real world entails a transformation of the third context—the doing of the exercise. Whereas before the student was to work alone, cloistered in his or her dorm room, now active participation with other students and faculty is the desired norm. Negotiation of constraints and performance specifications is best done in a recitation section of no more than a few dozen students. Encouraging students to team up in pairs or groups of three or four to discuss patient weights, bump heights, bed dimensions, and possible constraints, enables learning. Since there are no "right" answers, students are less restrained in speaking out and voicing different estimates or conjectured constraints.

We ask students to keep journals of their work as individuals. The design of a hospital bed, or even just its wheels, could take a full semester—but we constrain the task to the sizing of the diameter of the wheels and ask the students to complete their analysis in one night. They are told to hand in their journal the next day in lecture. We take care to ensure that by the end of the initial encounter in recitation, the students have a good grasp of the remaining task.

That remaining task includes isolating and applying the equations of static equilibrium to produce the necessary relationships among step height, patient weight, bed dimensions, push force, and wheel diameter. Again, from a traditional perspective one might claim that nothing has changed; but even here, the context for doing is significantly different. Recall that in the roller problem the author asked for a specific answer—the magnitude of the pushing force—and provided all the information needed by the student to produce this result. No extraneous, or diverting, information appeared in the problem statement. Now, with multiple responses possible and with uncertainties associated with the ranges of values for the parameters which appear in the equations derived from the principles of static equilibrium, "solution" of the task becomes less straightforward. In fact, a spreadsheet might be profitably employed to explore the sensitivity of wheel diameter to bump height, push force, ranges of patient weights, and bed dimensions. The nature of the solution process changes in significant ways.

It is this process, as recorded in the journal, that is the basis for evaluation of the student's work. Now, in contrast to the roller problem, students are encouraged to make note of the resources they call upon—fellow students, textbook tables, even a call to a supplier. They are asked to explicitly state their assumptions (about bumps, attendant's pushing force, and the like), performance goals, and design constraints. They must learn to summarize the output of a spreadsheet analysis concisely for the reader, and articulate and defend their approach and results.

This description of a day's exercise in an engineering mechanics course is meant to illustrate the nature of the transformation in undergraduate engineering education we consider necessary. When we view this transformation within the abstract world of the problem, the big world "out there," and the context of the classroom, both form and content change. It is not just a matter of adding another required course; in fact, nothing need change in this respect. It is, rather, a matter of changing the process and culture of engineering education, and of seeing our objectives and tasks from a different perspective—one that acknowledges the importance of context, the presence of ambiguity and uncertainty, and the need to exercise judgment and make efficient use of a broader spectrum of resources. It encourages students to work in teams, articulate their ideas clearly, and defend their approaches.

There are, of course, barriers to significant change. Some will argue that the new exercise will take significantly more time than the old. This is true; but we ask if the traditional accounting procedures invoked in curriculum planning discussions make sense. Does assigning more problems mean more learning? In this case, we think not.

The new method will certainly take more time in the sense that faculty need to learn how to manage and teach in this new mode. Resource materials need to be developed. Computer tools must be readied and made available for student use. Evaluating student work as expressed in a journal will take more time, at least initially. Most important, faculty will be asked to acclimate themselves to a new culture—one that calls on them to teach in a way that they were not taught. The transition from a lecturer and evaluator of problem sets filled with single-answer puzzles, to a more contingent and open-ended relationship with students will pose an enormous challenge to most faculty. But as time goes on and experience accumulates, as student patterns of response become apparent, and as faculty become more adept, the new will seem more familiar.

In one other very important way, the transformation embodied in the new form of this exercise calls for change: it strongly suggests that the traditional criteria for faculty appointment, tenuring, and promotion no longer suffice. Expertise in a single, specialized research discipline is no longer enough. What is required is more general knowledge of engineering practice, coupled with expertise in a traditional field or subdiscipline. Faculty are no longer simply fonts of knowledge, with students playing the role of the sink. We need new images of the teacher. "Coach" is one; "mentor" another.[6] But, like good teaching, good coaching is not just a matter of form, independent of content. To be able to judge when a student's question is pure folly, potentially significant with some slight shift of context, or right on target requires technical expertise. It also requires experience of engineering practice.

Defenders of the present system may protest that an engineering education must focus on the latest, most up to date engineering knowledge and incorporate the widely accepted findings of current research. We agree, but challenge the

6. For an excellent discussion of coaching as a model for professional education, see Donald A. Schön, *Educating the Reflective Practitioner* (New York: Basic Books, 1988).

boundaries of engineering research and engineering knowledge as they are currently drawn. A broader, more inclusive definition, one which incorporates research supporting improvements in the process world as well as the growth of object world thinking, would, we believe, do more to advance our thinking and improve industrial effectiveness than the current narrow focus on science alone. In addition to research in engineering science, why not research engineering processes, the relationships between consumer use and design features, the management of dispersed design, the social impact of technology, or the social and technical context of alternative technology?[7]

We believe that many benefits follow from acknowledging that there are in fact two worlds of engineering work. First and most obvious, it gives us a way to think about how engineering education must change in order to better prepare engineering graduates for the real work that engineers do. In a time when our national focus was on military and space development, and American political and economic dominance virtually assured a market for American consumer products, engineering education that prepared students for object world work alone was adequate, if not ideal. As economic and political realities change, however, few engineering graduates can afford to be unprepared for the social process dimension of engineering practice.

A second benefit of the changes we recommend in engineering education deserves mention. Echoing Dean Streett, we note that the tendency toward exclusion of women and minorities (African Americans in particular) from engineering represents a loss to the innovative capacity of the country and a loss to individuals as well. Changes that retain the fundamental character of engineering practice and the high standards prevailing in the field while also making it more

7. Broadening the definition of engineering research can help to enliven undergraduate education. If *process* itself becomes a subject of research and of reflection by faculty, then the approach as well as the content of undergraduate teaching may be positively affected. The classroom could become a laboratory for understanding teamwork and handling competing interests, negotiation, and other process-related subjects. Presently, the purpose of undergraduate education is to convey scientific fundamentals, and this can easily become repetitive and tedious for research-oriented faculty. A wider definition of what constitutes legitimate and useful engineering research would improve this situation. In his excellent and influential report, "Scholarship Reconsidered: Priorities of the Professoriate," Ernest Boyer (1990) proposed that we expand the definition of valued faculty activity to include not only what is commonly known as "research," but other forms of scholarly activity as well. In Boyer's lexicon, standard research becomes "the scholarship of discovery," while three other forms of scholarship are also recognized: the "scholarship of integration," the "scholarship of application," and the "scholarship of teaching." In most universities, only the first is highly valued and consistently rewarded, despite the fact that the other forms of scholarship also make important contributions and are very attractive to faculty. All of these forms of scholarship can be conducted in a committed, rigorous way that advances human knowledge and capability, and Boyer argues that all should therefore be recognized and rewarded. An engineering school that adopted this expanded definition of scholarly activity would be opening the door to a creative rethinking of what is valuable and effective in engineering education.

attractive to underrepresented groups would not only bring new points of view into engineering practice, but would meet important social goals as well. We speculate that perhaps women, African Americans, and members of other underrepresented groups are attracted to engineering because of its pragmatic, creative, and essentially optimistic character, but are turned off by engineering education's preoccupation with object world concerns and relative de-emphasis of process and other contextual issues, including social and ethical concerns. A curriculum emphasizing process, context, ambiguity, and the importance of the social world could well attract new students who have been wary of the strictures of the current practice of engineering education. We believe that engineers from diverse backgrounds, broadly educated along the lines we have suggested, would enrich the field of engineering and make an enormous contribution to our nation's productive capacity.

10

# THE SENSELESS SUBMERGENCE OF DIFFERENCE: ENGINEERS, THEIR WORK, AND THEIR CAREERS

*Leslie Perlow and Lotte Bailyn*

Definitions of engineers' work which cast the engineer as "interested primarily in the application of scientific knowledge about the natural world and in discovering facts about the artificial world created by humans" (Junior Engineering Technical Society, 1989:3) sit side-by-side with definitions of the successful engineering career as moving away from technical work into management. As Everett Hughes said almost forty years ago, "the engineer who, at forty, can still use a slide rule or logarithmic table, and make a true drawing, is a failure" (1958:137). These received notions about the nature of engineering work and the definition of the successful engineering career presume that engineers' work and careers are much more homogeneous than they really are. Definitions of engineering as merely the "application of scientific knowledge" deny the actuality of engineering practice, as vividly described in the preceding chapter, and the different preferences that engineers bring to their work. Similarly, the "failure" implied by a continuation of technical work ignores the variety of engineers' career orientations, and the many different roles necessary for successful engineering work in industry.

The tension between this educationally and occupationally based definition of engineering work on the one hand, and the organizationally based definition of career success on the other, has been the core of much previous work on engineers (e.g. Van Maanen and Barley 1984; Raelin 1985; Whalley and Barley, Chapter 1). The assumption underlying these analyses is that all engineers are in agreement on what defines engineering work and what constitutes career success, which leads the research on engineers to focus on "the" engineer, "the"

We are grateful to Stephen Barley, Sarah Kuhn, and John Van Maanen for perceptive comments on previous drafts of this paper.

work of the engineer, and "the" engineering career (e.g. Perrucci and Gerstl 1969; Ritti 1971; Zussman 1985; Whalley 1986b; Kunda 1992). A picture has emerged of the "generic" engineer, the "generic" engineering job, and the "generic" engineering career.

The tacit assumption underlying this picture is that engineers are essentially a homogeneous group. Such a perspective acknowledges no differences among engineers in the types of contributions they wish to make or the career paths they would like to pursue. Engineers do not, however, possess a unified set of values, preferences, and career orientations—and organizational goals require them to perform a wide range of occupational activities.

Engineers bring individual experiences and preferences to their employment. This heterogeneity, however, is often submerged because both the occupational community's definition of the valued work of an engineer and the organization's definition of the opportunity structure in which that work takes place are excessively narrow and to a great degree homogenizing. They value one particular way of working, to the exclusion of other activities (e.g. activities in the social world), and value one particular way of having a successful career.

In this chapter, our intention is to indicate that these definitions of valued work and successful careers ignore important distinctions in what engineers bring with them to the workplace and what organizations actually require from them. The thesis of our paper is that a *latent congruence* between engineers' preferences and organizational activities exists, but that the potential advantage of this congruence remains untapped because of the acceptance of monolithic definitions of work and career which serve to submerge existing and potentially valuable differences among individuals and their roles and activities.

## ENGINEERING WORK

Much of the data presented in this section stem from our current work studying engineers in a large, successful high-tech company.[1] Along with our colleague Joyce Fletcher, we spent three years involved in an intensive study of design engineers in a product development division of this company. A contrast between the engineer's definition of work and the organization's actual requirements for that work emerged from this field research.

The data we use are based on an initial project focusing on mechanical engineers which included interviewing, observing, and shadowing engineers, man-

1. This study has been supported by the Ford Foundation under grant # 910-1036.

2. During this period, we conducted formal interviews lasting one to two hours each with thirty-eight engineers, sixteen managers, and twelve members of the support staff; attended a number of technical and managerial meetings; and shadowed a few engineers and one manager throughout their day. Some of this fieldwork was done by Maureen Harvey, another member of our research team.

agers, and support staff;[2] on Fletcher's analysis of invisible work, which was based on her study of the work of six women engineers (Fletcher 1994); and on a nine-month study by Leslie Perlow which tracked the work lives of seventeen software design engineers throughout the life cycle of a project, from the commitment of funding to product launch (Perlow 1995).[3]

In our initial phase, we asked engineers, "What do you do?" In response, they tended to contrast what they labeled "real engineering" with everything else that took time away from these activities. According to these engineers, real engineering is the technical component of their assigned work. The essence of real engineering, they told us, is analytical thinking, mathematical modeling, and conceptual problem solving. As one engineer described it, real engineering is "designing, sitting at a desk or a computer, using the laws of nature to create something." Real engineering is work requiring scientific principles, independent creativity, and the skills that one acquired in school. One engineer summed it up as follows: "Real engineering is what I thought I was hired to do."

Real engineering tended most often to be defined in contrast to whatever was keeping engineers from doing such work. As one engineer said, "managing, coordinating, administering, helping . . . these activities are not real engineering." Another commented, "The biggest frustration of my job is always having to help others and not getting my own work done." Still another complained, "Now that I manage two people, this leaves me without a job. I spend my time telling them what to do, statusing things, and going to meetings."

Underlying this emphasis on technical problem solving, however, we found a range of diverse tasks required to meet the needs of the engineering job. Louis Bucciarelli and Sarah Kuhn, in the preceding chapter, discuss the mismatch between the content, structure, aims, and values of contemporary engineering education—focused on object world thinking—and what is actually required of an engineer to function effectively in today's industrial settings. They claim that the formal education process teaches engineers to succeed in the object world, but overlooks the large social components of discussion, negotiation, exchange, integration, and consensus building that are a fundamental part of the design process.

In our research, we distinguish between "real engineering" and the "rest of the job." It is the rest of the job that provides the "glue" that makes it possible for the technical work of each individual engineer to hold together in a larger project. We have noted elsewhere (Perlow 1993) that the rest of the job consists of both "necessary evils" (i.e. meetings and paperwork) and "invisible activities" (i.e. behind-the-scenes work to facilitate the functioning of the project).

3. This research included formally interviewing the seventeen engineers as well as having them track on five different randomly selected days over the course of five months what they did from waking up in the morning until going to bed at night. These daily trackings were followed the next day by a detailed debriefing interview.

These invisible activities have been analyzed in detail by Fletcher (1994). She identified four general sets of activities that are necessary for project success but which go widely unrecognized ("get disappeared," to use her term) in a culture that only rewards technical output. The first set she labels *preserving*: activities, such as passing on necessary information or creating bridges between people, that are associated with preserving the life and well-being of the overall project, independent of one's own deliverables. The second group she calls *empowering*: activities related to empowering others to achieve and contribute to the project. Third are *achieving* activities which empower oneself to develop and grow in order to achieve goals and contribute to the overall project. The fourth set she calls *creating team*: activities intended to create the social entity of a team (see Fletcher 1994, 78).

These activities constitute work in what Bucciarelli and Kuhn have called the social world, whereas real engineering is the work done in the object world. As they make clear in their chapter, however, these worlds are not distinct. Our data support their argument that in design the object world and the social world are fundamentally intertwined. Engineering is not just technical problem solving, nor is it technical problem solving on the one hand and integration, negotiation, communication, and collaboration on the other. Rather, performing the design job requires a merging of the two worlds. Consider the following example.

Two members of the software development team we studied each had their own set of technical deliverables. One, a software engineer, was responsible for a major part of the software design. The other, a mechanical engineer by training, was responsible for the integration of the software and the hardware. In order for the product to function, they depended on each other. Unless the software was functional, it was not possible to integrate the software and the hardware. However, without understanding the specifications of the hardware, one could not provide the necessary software.

One morning, three months into a nine-month development cycle and five days before an important release requiring the integration of the software and the hardware, Steve, the engineer responsible for the integration, came into the lab where Andy, the software designer, worked, and requested the computer that Andy was using. Andy described the situation as follows: "Steve showed up and demanded that—well he didn't exactly demand—but basically he needed me to help him with some stuff . . . He needed me to be doing some stuff at the machine to help him. It would have taken him an awful lot longer by himself . . . Steve's interruption was of great value to him and little value to me . . . It was worthwhile for him because it allowed him to go on with his work." When Andy used his technical skills to help facilitate Steve's work, Andy perceived it as supporting Steve's real engineering but not his own. From Andy's perspective, helping Steve was an activity in the social world; but it was Andy's technical skills that enabled him to help Steve in this way.

This example demonstrates the relationship between the social and object worlds and shows how engineering work can span the boundary between them. It further shows how an activity that is critical to the development of a product is not considered real engineering. Although Andy was aware that Steve needed his help in order to complete the integration, he still considered Steve an interruption that took time away from his own work. In this way, the narrow definition of what constitutes real engineering can interfere with the successful and timely completion of a technical project.

It is clear that successful product design requires activities beyond those that are considered real engineering. Furthermore, real engineering is not the only work that engineers like to do. Despite the fact that engineers are trained to do real engineering and not the rest of their jobs, and despite the fact that engineers proclaim real engineering to be the essence of their jobs even though there are other activities equally necessary, not all engineers prefer real engineering all of the time.

While engineers may appear to conform, when they are questioned about their work preferences and career orientations, no underlying consensus is found. For example, as part of the initial phase of our research project, Perlow (1993) did interviews and Q-sorts with design engineers at the same level and with nearly identical responsibilities, in order to examine their preferences for different work activities. From this analysis, two ideal types of preferences emerged: a preference for technical problem solving and a preference for the interactive work necessary to facilitate the engineering project (Perlow 1993).[4]

One engineer we interviewed was Carol, an individual with a preference for solving technical problems. According to Carol, after college she "lucked out and got a typical design engineering position—with real hands on lab work." This was her preference, but, unfortunately for her, within six months her job changed to working on "the reliability aspect of the project" which she did not value. Two years later she had new responsibilities on the project, but, as she explained, "I am still stuck. It hasn't changed. I haven't been able to shake the role. I am still the reliability expert . . . People constantly call for help with this part. I hate it." Carol preferred solving her own technical problems, not helping others with theirs. As she said, "I used to want to go into management, but now that I am realizing that interpersonal skills are the most important part of management . . . I no longer am so sure. Instead, I am considering computer science . . . then, I can be an individual contributor."

4. We are not implying that people always fit into one or the other category; we are only indicating that varying preferences exist. Furthermore, people's preferences are negotiated and may change with time and circumstance.

Kim, on the other hand, was an individual with a preference for interaction. During the interview Kim spent little time talking about her own engineering work and much time describing the work environment. "It isn't bad that everyone doesn't do the real engineering. If they did, who would do the work? Someone needs to do the other stuff. It is a different matter to make something a practicality. I like the brainstorming and the coming up with ideas, but I also like making it work." In further describing her skills, Kim said, "I have been told that I am good at getting a given task done—even without knowledge of how to solve it. I jump in and fix it . . . Also, I am good at working with people and getting them to do what I need." Kim added, "I think of myself as a very directive person . . . directing designers and technicians and planning and leading meetings, those are things I like to do."

Both of these engineers considered themselves to be "real design engineers" and desired to do the associated technical work. Their different preferences, however, highlight the various ways that engineers respond to their work. But the engineers we studied were not encouraged to express these preferences; rather, they perpetuated the myth that they preferred real engineering (Perlow 1993).[5] In the end, both the engineers and the project lose when the value of these other activities is not recognized.

A vivid example comes from the software design group we studied. In this group, there was one male engineer without whom the project would have failed, according to both his peer engineers and his managers. He was not only the technical guru, but the group's source of technical support. Yet, he was often punished for the time he spent helping others to get their work done. According to this engineer, "Sometimes others have a question, or just need a boost, and many people around here just say no. I don't think they mean anything bad by it. I just think that they are the type of people who can say no, whereas I am not." He then went on to add, "As a result, I don't get as much done around here." It was on this latter point that his managers focused their attention. No one acknowledged that the helping behavior demonstrated by this engineer was real work. Rather, this engineer was told repeatedly by his managers that he had to emphasize his own work, and that if he wished to spend time helping others he would have to do it on his own time. The reward system was based on individuals accomplishing their real work—their own technical deliverables—not facilitating anyone else's work, even for the project's benefit.

The emphasis on technical problem solving, as opposed to behind-the-scenes work, forces engineers like the one described to reduce the time spent helping others in order to produce more of their own output. Yet, this group would not

5. In fact, some of the engineers whose preferences were for interaction were themselves surprised when they realized that their preferences did not match their own definitions of real engineering.

have succeeded if it had not been for some engineers' willingness to help others, even at the expense of their own work and therefore their own recognition.

Another instance where "invisible activities" were critical but not rewarded was brought up at a presentation to managers at the high-tech company we studied. When presented with the finding that their engineers perceived individual-centered technical contributions to be more important than team-centered, supportive behaviors, most of the managers recognized that they had a problem. They agreed that it takes more than a collection of individual contributors to create a complex product—yet, they only knew how to reward output, and therefore did not know what to do. One manager recalled his own experience trying to promote an engineer who had provided the "glue" for his entire team. This manager had run into difficulty when he tried to explain to his management why this particular engineer should be promoted. The engineer had been critical to the success of the product, but there was no tangible individual accomplishment to which to point, and the promotion did not go through.

By failing to recognize and value the whole job of the engineer, both the engineers who prefer activities not constituting real engineering and the organization that requires this very diversity in order to develop a successful product lose out.

## ENGINEERING CAREERS

A similar argument, we think, can be made for engineering careers.[6] Hughes' claim that "the engineer who, at forty, can still use a slide rule . . . is a failure" resounds throughout subsequent research on engineers' careers (see e.g. Dalton and Thompson 1986; Katz 1988; Von Glinow 1988): a successful engineering career is organizationally defined as moving away from technical work and into management. Because engineering is characteristically performed in large organizations, it shares with all organizational careers the fact that hierarchical movement provides the cultural definition of success in that setting. To become a manager is to be successful; remaining an engineer typically means less pay, less status, and less influence. This narrow definition, however, ignores the multiplicity of roles necessary in technical organizations and the variety of career orientations that exist among technical personnel. As in the case of engineering

6. The data on which this section is based stem both from our current research project as well as from our previous studies of technical careers (e.g. Bailyn 1982, 1985, 1991; Bailyn and Lynch 1983).

work, a restrictive definition of career success and the presumed homogeneity on which it is based work against both individual and organizational goals.

Some organizations have recognized this dilemma, and provide engineers with the alternative of staying on a technical track. In theory, such a career route is meant to provide the same movement in pay and status that is available on the managerial career path, but allows the engineer to continue with technical work. The reality, however, is different. People choosing this route tend to work individually in narrowly specialized areas. Rarely are they assigned to projects that present them with new challenges as opposed to projects for which they already possess the necessary specialized skills. Nor is movement on the technical ladder accompanied, usually, by any increase in influence over technical decisions (cf. Epstein 1986). Thus engineers who choose the technical ladder have little opportunity for new learning in either the object or the social world (see also Gunz 1980; Allen and Katz 1986; Katz, Tushman, and Allen 1992). In most companies, technically oriented engineers sooner or later face stagnation—the feeling of being stuck with no opportunities for professional or technical development. The dilemma was well expressed by one employee of an R&D lab, who said, "Everyone wants it [to be a manager]. But when one looks at management I wonder why. What seems to happen is that at every level the only thing you want is the next level" (Bailyn 1991, 3). The definition of the successful engineering career as one that inevitably moves into management, though perpetuated and reinforced by organizational norms and practices, is actually counterproductive because it denies the multiple roles necessary in a technical organization.

Bringing a product to market requires a variety of critical roles, such as designing, selling (both internally and externally), and coordinating resources (Schriesheim et al. 1977; Roberts and Fusfeld 1982; Bailyn 1985, 1991). For example, to ensure profitable innovation, there has to be a successful transfer of technology from research and development labs to operations. Typically, however, such a role is not explicitly recognized or valued—and yet, as will be seen, there are technical workers in these labs who actually prefer such roles. Thus the labs, by failing to recognize this heterogeneity, lose the advantage of the very diversity they need. Another organizational difficulty produced by this narrow definition of the successful engineering career stems from the fact that individuals generally get promoted on the basis of their technical skills, and then deprive their subordinates by continuing to be overinvolved in the strictly technical aspects of the work. As the chief engineer in an R&D lab remarked about the newly appointed manager of his division, "You don't keep a dog and then bark yourself!" (Bailyn 1991, 4).

The restricted and narrow definition of career success leads to an undervaluation of the variety of roles actually critical for successful product design. Moreover, this same narrowness also overshadows the diversity of career orientations found among engineers. James M. Watson and Peter F. Meiksins (1991), for example, used surveys, mailed to 800 engineers in the Rochester area, to address the question, "To what extent do engineers have professional or organizational values?" (p. 145). They concluded that "work values of our sample of engineers are varied. It is almost impossible to see them as a homogeneous group, or even to develop a viable dichotomized model of their work values" (p. 153).

Beyond a diversity in work values, there is a desire among engineers for more variability in their career paths. Despite common educational backgrounds, despite similar early career interests and experiences, and even despite consensus on what constitutes real engineering, engineers' orientations at mid-career take a number of different forms (Bailyn 1982). For example, from detailed analysis of interviews with sixteen technical staff members in a corporate R&D lab, five different orientations emerged, each of which was accompanied by different reactions to work experiences and by positive and negative feelings about different aspects of the lab's career procedures (Bailyn 1985). Five people were oriented to "real engineering." They were concerned with technical craftsmanship, with the development of a product or process that could, in some way, be identified with themselves. In this lab, however, these employees were not the best rewarded nor the most satisfied. Rather, two professionals more oriented to adding to knowledge than to contributing to a particular product were the most satisfied. Three individuals who were oriented to technical management and desired a combination of technical work and managerial responsibility and authority were the most highly rewarded—but they were not the most satisfied.

Two other orientations, important for the lab's success, were present but essentially unrecognized, a situation that led to dissatisfaction among their proponents. Three engineers were oriented to production, and hence deeply committed to the follow-through on R&D. They worked hard to ensure that the lab's work would get translated into profit for the corporation, but their contribution tended not to be valued. Another three were oriented to administration. They were less involved with the technical side of their work and more with the lab's administrative tasks, such as the scheduling and budget control of projects, and the evaluation and development of people. In some ways they were the perfect embodiment of the successful manager in an engineering setting, but because they no longer excelled in the technical part of the work—in the real engineering—they too were undervalued.

These responses reflect the strategic thrust of this lab—its particular definition of valued work and career success. But the lack of correspondence between satisfaction and rewards—a common result of presuming homogeneity where it does not exist—alerts us to a final contradiction. In engineering, the definition of what career path is seen as most successful conflicts with the work that is most valued. There is a conflict between the occupational definition of engineering as consisting only of real engineering, and the organizational definition of hierarchical movement into management as constituting career success. The occupation defines technical problem solving as high status, valuable work, but the organization promotes a definition of the successful career path as one that leads engineers away from this activity.

Engineers find themselves torn between real engineering and career success. The engineers we studied spoke of the decision of whether or not to go into management as a critical choice in their lives. They talked about the tension between doing what they had been trained to do—the work that gave them internal satisfaction—and doing what would make them successful in the organization. We also found managers ambivalent about their positions. Some expressed insecurity about their personal contributions. They regretted their lack of tangible results and lamented no longer having technical deliverables—no longer being able to do the real engineering. They expressed a sense of worthlessness, since they could no longer point to a unique contribution that they had made. Since their contribution was invisible in the final product, they found it harder to identify with it.

This personal dilemma was well expressed by an engineer in a previous study (Bailyn and Lynch 1983). Speaking about engineering from the vantage point of having been promoted into management, he commented, "One of the advantages in engineering is that you can see the product of your efforts, something tangible that you were responsible for, some concrete contribution that can make you feel proud of yourself . . . In engineering you are quickly calibrated or evaluated. You can measure the efficiency of a product and this gives you an evaluation of your own success or failure. . . . That's the advantage of hard core engineering" (Bailyn and Lynch 1983, 280). He found none of this in his managerial position—but the pull of organizational success was strong enough for this man not to want to return to engineering; he wondered, rather, whether he should have gone into another profession altogether.

We are obviously not the first to note this contradiction. John Van Maanen and Stephen R. Barley (1984), for example, argued that engineers experience an organizational-occupational tension, and Joseph Raelin (1985) described the situation as a "clash of cultures." By referring to engineers as corporate professionals, Whalley and Barley (Chapter 1) give another name to this dilemma. These approaches, however, do not deal with the underlying issue of presumed homogeneity. Our argument is that the seeming contradiction between work and career—between occupational and organizational pulls—is, in part, a function of this neglect. It is the monolithic view of engineering work that clashes with the monolithic view of the successful engineering career path. What this chapter attempts to demonstrate is that such a clash is not inevitable; rather, it arises from a lack of recognition of the differences that exist among individual engineers and among the activities that make up their jobs.

## GENDER IMPLICATIONS

We have seen that the work preferences and career orientations of engineers span a much wider spectrum than is encompassed in the socially constructed definitions of real engineering and the successful engineering career. These definitions, based as they are on a presumption of homogeneity, will naturally fit some people better than others—and since technical work and the shape of the engineering career have been largely defined by men, they are more likely to fit easily with men's lives and with men's skills and styles. Both of these restricted definitions, therefore, are likely to make the role of women in engineering problematic.

It has been found, for example, that women engineers more frequently prefer interaction (e.g. Perlow 1994). And, as Judith McIlwee and J. Gregg Robinson (1992) note, "People skills do not pay off the way they should, because it is male engineers who define the criteria for the good engineer . . . Where male engineers are most powerful, being female exacts the greatest price" (p. 141). Similarly, women have been found to be more likely to do the "invisible" work (Fletcher 1994).

Since the culture of engineering privileges technical skills at the expense of other necessary skills such as facilitating, coordinating, and integrating, it benefits those (primarily men) who are likely to prefer these activities, and hinders those (primarily women) whose preferences are likely to be elsewhere (cf. Bailyn 1987). As Fletcher (1994) has documented in detail, the kinds of behaviors that women engineers engage in—e.g. problem prevention, team coordination, or, in general, relational activities—though critical for the success of the project, tend to "disappear" in an individually oriented male culture. Thus, the narrow definition of real engineering and the resulting submergence of individual differences may have a particularly negative effect on women engineers, and hamper the goal of gender equity in technical work.

Women may also be undermined by the definition of career success as hierarchical movement. Underlying this definition is the presumption that work must at all times be the top priority in everyone's life—otherwise one is not seen as worthy of promotion. Since women are still more likely to have to (or want to) take responsibility for the "second shift" (Hochschild 1989), this presumption is particularly problematic for them. If work and family are considered competing priorities, not compatible or synergistic responsibilities (Perlow 1994; Andrews and Bailyn 1993), women will find it difficult. As one female engineer put it: "I really like my job. I don't want to give it up but I will. I know other women who are doing the same thing, leaving the field, that is." Thus the restrictive definition of career success as hierarchical movement with its implication that work must always come first may lose for the organization the talents of people concerned predominantly about their families and communities; and it is in those arenas, rather than in school or on the job, that work in the social world may best be learned.

Both of these constraining definitions, therefore, are likely to create conflict for women engineers, and may prevent organizations from tapping their full potential.

## REDEFINING REAL ENGINEERING AND CAREER SUCCESS

In this chapter, we have argued that engineering work and engineering careers are defined much more narrowly than actually represents the needs of the projects on which engineers work. We have also shown that engineers do not all prefer to do real engineering all of the time and that they do desire more flexible and imaginative career paths. We are now left to ponder the lack of congruence between individual differences and job requirements.

If engineers had preferences for work activities and career paths that were not required for organizational success, or if there were tasks and roles required that engineers preferred not to do, then a tension between requirements and preferences would clearly exist and would need to be addressed. However, when individual differences are brought into the picture, it turns out that there is sufficient variety among individual preferences and orientations to meet the demands of the engineering job—but such a congruence will never become manifest as long as organizations and occupations continue to perpetuate norms of homogeneity. The seemingly senseless submergence of differences stems from an overly constrained, monolithic view of both job requirements and preferences.

There exists then a latent congruence between tasks and preferences that needs to be addressed, rather than limiting the discussion, as previous authors have done (e.g. Van Maanen and Barley 1984; Raelin 1985), to the incongruence that appears on the surface. Managers and engineers need to come to terms with the discrepancy between real engineering and the day-to-day needs of the engineering job. Engineers uniformly subscribe to the idea that real engineering is the essence of their work—and yet, real engineering is only one aspect of the job of the engineer and it is not always preferred by all engineers.

The problem stems from the fact that real engineering is the work we know how to value; it is about individually solving technical problems, as opposed to solving problems through interaction with other people. In our society we know how to value outputs; we do not know how to recognize the contribution of activities necessary for the creation of the product but for which there is no tangible output (cf. Thomas 1994; Bucciarelli and Kuhn, Chapter 9). We need a language, concepts, and a framework with which to understand, recognize, and value such non-output-oriented work (cf. Perlow 1993, Fletcher 1994). Developing this new and more inclusive definition of what is valued in engineering work will not be easy because it goes against our individualistic way of thinking about competence and rewards. To be effective, such change will have to be systemic; it will take a long time and require individuals to work together.

Equally limiting is a definition of career based on the assumption that success means movement up a managerial hierarchy. As the situation stands today, engineers who want to continue in the technical field are frustrated by companies that do not provide them with a career plan that employs their talents and leads to recognition of their work. Nor do organizations necessarily benefit when they promote their best engineers to management. Engineers who are promoted to supervisory roles damage themselves and their subordinates if they remain so involved with the details of technical projects that they deny their subordinates the right to their own work.

What is needed, therefore, are new and more inclusive definitions of work and of career. Careers need to encompass more than just hierarchical position. They must be seen as a series of work tasks, akin to the sociological definition of career. These tasks cannot be limited only to the narrow occupational definition of real engineering but must include the rest of the job as well. Under these conditions, management, rather than being a status to be attained, becomes a task to be performed, and people can go back and forth among the various tasks of the engineering job (seen in its broadest sense) without either failure or success being attributed to such moves.[7]

Under such broadened definitions, work and career could come together in more diverse ways—in what has been suggested as the hybrid career (Bailyn 1991). Such a career consists of multiple work assignments, each with different forms of evaluation and reward, strung together in a series of discrete and discontinuous segments. Even though the technical work force is, almost by definition, specialized, the assumption behind multiple work assignments is that overspecialization must be avoided. Individual career segments would be planned according to the particular combination of individual and organizational needs at that point in time. Subsequent segments would start the planning from scratch—

7. We are aware that this view ignores the power implications behind the differentiation of management from non-management. Nonetheless, we think that future close-grained studies of the nature of managerial work—the tasks actually involved—could help us unpack this seemingly tight correlation and make the enactment of this idea a realistic possibility.

what we might call zero-based career planning. All tasks, technical and interactional, would be recognized as valuable. The hybrid career would allow people to move easily among multiple tasks, and the resulting discontinuities, because preplanned, would not be seen as failures. In this way, organizations could take advantage of the diversity that exists among their technical employees, and engineers could freely express their inclinations and talents in productive work.

Current definitions of work and career clash because the organizational definition of career success emphasizes hierarchical position, and the occupational definition of what is real engineering is limited to the strictly technical aspects of the engineer's job. Since engineer's preferences are in fact much broader than these narrow definitions, and since organizational and project requirements extend beyond the scope of these narrowly defined terrains, the clash seems unnecessary. Our intention has been to show that benefits are lost to engineers personally as well as to their employing organizations as a result. When we consider the technical worker of the future, we need to probe beneath homogeneous constructions and develop ways of recognizing rather than submerging existing differences.

# REFERENCES


Abbott, Andrew. 1988. *The System of Professions.* Chicago: University of Chicago Press.

——. 1989. "The New Occupational Structure: What Are the Questions?" *Work and Occupations,* 16:273–91.

——. 1991. "The Future of Professions: Occupation and Expertise in the Age of Organization." Pp. 17–42 in *Research in the Sociology of Organizations,* vol. 8, edited by Pamela S. Tolbert and Stephen R. Barley. Greenwich, Conn.: JAI Press.

Adler, Paul. 1992. *Technology and the Future of Work.* New York: Oxford University Press.

Ahlstrom, Goran. 1982. *Engineers and Industrial Growth: Higher Technical Education and the Engineering Profession during the Nineteenth and Early Twentieth Centuries: France, Germany, Sweden, and England.* London: Croom Helm.

Allen, Thomas J., and Ralph Katz. 1986. "The Dual Ladder: Motivational Solution or Managerial Delusion?" *R&D Management,* 16:185–97.

American Academy of Orthopaedic Surgeons. 1987. "Emergency Care and Transport," 4th ed. Park Ridge, Ill.: American Academy of Orthopaedic Surgeons.

American College of Emergency Physicians. 1984. "Medical Control of Emergency Medical Services." Dallas: ACEP.

Anderson, Margo. 1994. "Only White Men Have Class: Reflections on Early Nineteenth Century Occupational Classification Systems." *Work and Occupations,* 21:5–32.

Andrews, Amy, and Lotte Bailyn. 1993. "Segmentation and Synergy: Two Models of Linking Work and Family." Pp. 262–75 in *Men, Work and Family,* edited by Jane C. Hood. Newbury Park, Calif.: Sage.

Applebaum, Herbert. 1992. *The Concept of Work: Ancient, Medieval, and Modern.* Albany, N.Y.: State University of New York Press.

Attewell, Paul. 1987. "The De-skilling Controversy." *Work and Occupations,* 14:323–46.

——. 1990. "What Is Skill?" *Work and Occupations,* 17:422–48.

Bailyn, Lotte, 1982. "Trained as Engineers: Issues for the Management of Technical Personnel in Mid-Career." Pp. 35–49 in *Career Issues for Human Resource Management,* edited by Ralph Katz. Englewood Cliffs, N.J.: Prentice-Hall.

——. 1985. "Autonomy in the Industrial R&D Lab." *Human Resource Management,* 24:129–46.

——. 1987. "Experiencing Technical Work: A Comparison of Male and Female Engineers." *Human Relations,* 40:299–312.

——. 1991. "The Hybrid Career: An Exploratory Study of Career Routes in R&D." *Journal of Engineering and Technology Management,* 8:1–14.

Bailyn, Lotte, and John T. Lynch. 1983. "Engineering as a Life-Long Career: Its Meaning, Its Satisfaction, Its Difficulties." *Journal of Occupational Behavior,* 4:263–283.

Baran, Barbara. 1987. "The Technological Transformation of White-Collar Work: A Case Study of the Insurance Industry." Pp. 25–62 in *Computer Chips and Paper Clips,* edited by Heidi I. Hartmann et al. Washington: National Academy Press.

Barkley, Katherine Traver. 1978. *The Ambulance: The Story of Emergency Transportation of Sick and Wounded through the Centuries.* Hicksville, N.Y.: Expedition Press.

Barley, Stephen R. 1984. "The Professional, the Semi-Professional and the Machine: The Social Implications of Computer Based Imaging in Radiology." Diss.: Massachusetts Institute of Technology.

——. 1986. "Technology as an Occasion for Structuring: Evidence from Observations of CT Scanners and the Social Order of Radiology Departments." *Administrative Science Quarterly,* 31:78–108.

——. 1988a. "Technology, Power, and the Social Organization of Work: Toward a Pragmatic Theory of Skilling and De-skilling." Pp. 33–80 in *Research in the Sociology of Organizations,* vol. 6, edited by N. DiTomaso and S. Bacharach. Greenwich, Conn.: JAI Press.

——. 1988b. "The Social Construction of a Machine: Ritual, Superstition, Magical Thinking and Other Pragmatic Responses to Running a CT Scanner." Pp. 497–539 in *Biomedicine Examined,* edited by M. Lock and D. R. Gordon. Dordrecht, The Netherlands: Kluwer Academic Press.

——. 1990. "The Alignment of Technology and Structure through Roles and Networks." *Administrative Science Quarterly,* 35:61–103.

——. "The New Crafts: On the Technization of the Workforce and the Occupationalization of Firms." Working Paper, Center for the Educational Quality of the Workforce, University of Pennsylvania, Philadelphia, 1991.

——. 1992. "Summary Report: Workshop on Technical Work." Report submitted to the U.S. Department of Labor. Program on Technology and Work, School of Industrial and Labor Relations, Cornell University.

——. "What Do Technicians Do?" Working Paper, Center for the Educational Quality of the Workforce, University of Pennsylvania, Philadelphia, 1993.

——. 1995. "The New World of Work." Washington: British North American Committee.

Barley, Stephen R., and Beth A. Bechky. 1994. "In the Backrooms of Science: The Work of Technicians in Science Labs." *Work and Occupations,* 21:85–126.

Barlow, Melvin L. 1967. "Educating the Technician." Pp. 408–26 in *History of Industrial Education in the United States,* edited by Melvin L. Barlow. Peoria, Ill.: Charles A. Bennett Co.

Becker, Gary S. 1975. *Human Capital.* New York: National Bureau of Economic Research.

Becker, Howard S. 1978. "Arts and Crafts." *American Journal of Sociology,* 83:862–89.

Beer, Ferdinand P., and Elwood R. Johnston. 1979. *Mechanics of Materials.* New York: McGraw Hill.

Bell, Daniel. 1973. *The Coming of Post-Industrial Society.* New York: Basic Books.

Benson, J. Kenneth. 1973. "The Analysis of Bureaucratic-Professional Conflict: Functional versus Dialectical Approaches." *Sociological Quarterly,* 14:376–94.

Berg, Ivar E. 1970. *Education and Jobs: The Great Training Robbery.* New York: Praeger.
Bishop, John H., and Shani Carter. 1991. "The Worsening Shortage of College-Graduate Workers." *Educational Evaluation and Policy Analysis,* 13:221–46.
Boland, Richard J. 1987. "The In-formation of Information Systems." Pp. 363–79 in *Critical Issues in Information Systems Research,* edited by Richard J. Boland and Rudy A. Hirschheim. New York: John Wiley.
Boltanski, Luc 1987. *The Making of a Class: Cadres in French Society.* Cambridge: Cambridge University Press.
Boyd, David R. 1983. "The History of Emergency Medical Services in the United States of America." Pp. 1–82 in *Systems Approach to Emergency Medical Care,* edited by D. R. Boyd, R. F. Edlick, and S. H. Micik. Norwalk, Conn.: Appleton-Century-Crofts.
Boyer, Ernest. 1990. *Scholarship Reconsidered: Priorities of the Professoriate.* Princeton, N.J.: The Carnegie Foundation for the Advancement of Teaching.
Brady, James T., Chester O. Gale, Richard C. Hodgson, Jeremy W. Johnson, William E. Moran, Andrew M. Rouse, Ralph Z. Sorenson, and Robert Stinson. 1959. *Teamwork in Technology/ Managing Technician Manpower.* Cambridge, Mass.: Harvard Business School.
Braverman, Harry. 1974. *Labor and Monopoly Capital: The Degradation of Work in the Twentieth Century.* New York: Monthly Labor Review Press.
Brint, Steven, and Jerome Karabel. 1989. *The Diverted Dream: Community Colleges and the Promise of Educational Opportunity in America, 1900–1985.* New York: Oxford University Press.
Brown, John Seely, and Paul Duguid. 1991. "Organizational Learning and Communities of Practice: Toward a Unified View of Working, Learning, and Innovation." *Organization Science,* 2:40–57.
Bucciarelli, Louis L. 1994. *Designing Engineers.* Cambridge, Mass.: MIT Press.
California State Department of Education. 1964. *Science and Engineering Technician Study: A Study of Employment and Education and Science and Engineering Technicians in San Mateo and Santa Clara Counties, California.* Sacramento: California State Department of Education.
Calvert, Monte A. 1967. *The Mechanical Engineer in America: 1830–1910.* Baltimore: Johns Hopkins University Press.
Cambrosio, Alberto, and Peter Keating. 1988. "Going Monoclonal: Art, Science, and Magic in the Day-to-Day Use of Hybridoma Technology." *Social Problems,* 35:244–60.
Carchedi, Guglielmo. 1977. *On the Economic Identification of Social Classes.* London: Routledge and Kegan Paul.
Carey, Max, and Alan Eck. 1984. "How Workers Get Their Training." *Occupational Outlook Quarterly,* Winter: 3–21.
Carnevale, Anthony, Leila Gainer, and Eric Schulz. 1990. *Training the Technical Workforce.* San Francisco: Jossey-Bass.
Cavendish, Ruth. 1982. *Women on the Line.* London: Routledge and Kegan Paul.
Center for the Study of Ethics in the Professions. 1982. *Beyond Whistleblowing: Defining Engineers' Responsibilities.* Chicago: Illinois Institute of Technology.
Chandler, Alfred D., Jr. 1977. *The Visible Hand: The Managerial Revolution in American Business.* Cambridge: Harvard University Press.
Child, John, and Janet Fulk. 1982. "Maintenance of Occupational Control: The Case of Professions." *Work and Occupations,* 9:155–92.
Clawson, Daniel. 1980. *Bureaucracy and the Labor Process.* New York: Monthly Review Press.

Collins, Harry M. 1974. "The TEA Set: Tacit Knowledge and Scientific Networks." *Science Studies*, 4:165–86.
——. 1990. *Artificial Experts: Social Knowledge and Intelligent Machines.* Cambridge, Mass.: MIT Press.
——. 1993. *Changing Order: Replication and Induction in Scientific Practice.* 2d ed. Chicago: University of Chicago Press.
Collins, Randall. 1979. *The Credential Society.* New York: Academic Press.
Coombs, R. W. 1984. "Innovation, Automation, and the Long-Wave Theory," Pp. 115–25 in *Long Waves in the World Economy,* edited by Christopher Freeman. London: Frances Pinter.
Crawford, Stephen. 1989. *Technical Workers in an Advanced Society: The Work, Careers and Politics of French Engineers.* Cambridge: Cambridge University Press.
Crompton, Rosemary, and Gareth Jones. 1984. *White Collar Proletariat: Deskilling and Gender in Clerical Work.* Philadelphia: Temple University Press.
Crosby, Philip B. 1985. *Quality without Tears.* New York: Plume.
Crozier, Michel. 1963. *The Bureaucratic Phenomenon.* Chicago: University of Chicago Press.
Curme, Michael, Barry Hirsch, and David MacPherson. 1990. "Union Membership and Contract Coverage in the United States, 1983–1988." *Industrial and Labor Relations Review,* 44:5–33.
Dalton, Gene W., and Paul H. Thompson. 1986. *Novations: Strategies for Career Management.* Glenview, Ill: Scott, Foresman.
Derber, Charles, ed. 1982. *Professionals as Workers: Mental Labor in Advanced Capitalism.* Boston: G. K. Hall and Company.
Diprete, Thomas A. 1988. "The Upgrading and Downgrading of Occupations: Status Redefinition vs. Deskilling as Alternative Theories of Change." *Social Forces,* 66: 725–46.
Doeringer, Peter B., and Michael J. Piore. 1971. *Internal Labor Markets and Manpower Analysis.* Lexington, Mass.: Ballinger.
Donnelly, J. F. 1991. "Science, Technology and Industrial Work in Britain, 1860–1930: Toward a New Synthesis." *Social History,* 16:191–201.
Edwards, Richard. 1979. *Contested Terrain.* New York: Basic Books.
Epstein, Karen A. 1986. "The Dual Ladder: Realities of Technically-Based Careers." Diss., Massachusetts Institute of Technology.
Evan, William. 1964. "On the Margin: The Engineering Technician." Pp. 83–112 in *The Human Shape of Work,* edited by Peter Berger. New York: Macmillan.
Fantasia, Rick. 1988. *Cultures of Solidarity.* Berkeley: University of California Press.
Ferguson, Eugene F. 1992. *Engineering and the Mind's Eye.* Cambridge, Mass.: MIT Press.
Field, A. J. 1980. "Industrialization and Skill Intensity: The Case of Massachusetts." *Journal of Human Resources,* 15:149–75.
Fletcher, Joyce. 1994. "Toward a Theory of Relational Practice in Organizations: A Feminist Reconstruction of 'Real' Work." Diss., Boston University.
Fletcher, W. Wendell. 1990. *Workplace Training: Competing in the New International Economy.* OTA-ITE-457. Washington: Government Publishing Office.
Franke, Walter, and Irvin Sobel. 1970. *The Shortage of Skilled and Technical Workers.* Lexington, Mass: Lexington Books.
Freeman, Richard B. 1976 *The Overeducated American.* New York: Academic Press.
Freidson, Eliot. 1973a. "Professions and the Occupational Principle." Pp. 8–29 in *The Professions and Their Prospects,* edited by Eliot Friedson. Beverly Hills, Calif.: Sage.
——. 1973b. *The Professions and Their Prospects.* 2d ed. Beverly Hills, Calif.: Sage.

——. 1977. "The Division of Labor as Social Interaction." Pp. 13–26 in *Work and Technology,* edited by Marie R. Haug and Jacques Dofney. Beverly Hills, Calif.: Sage.

——. 1986. *Professional Powers: A Study of Institutionalization of Formal Knowledge.* Chicago: University of Chicago Press.

Galbraith, John Kenneth. 1967. *The New Industrial State.* New York: Houghton Mifflin.

Giddens, Anthony. 1982. "Power, the Dialectic of Control and Class Structuration." Pp. 29–45 in *Social Class and the Division of Labour,* edited by Anthony Giddens and Gavin MacKenzie. Cambridge: Cambridge University Press.

Goldman, Alvin I. 1987. "Foundations of Social Epistemology." *Synthese,* 731:109–44.

Goldthorpe, John H. 1982. "On the Service Class, Its Formation and Future." Pp. 162–85 in *Social Class and the Division of Labour,* edited by Anthony Giddens and Gavin MacKenzie. Cambridge: Cambridge University Press.

Gorz, André. 1967. *Strategy for Labor.* Boston: Beacon Press.

——. 1972. "Technical Intelligence and the Capitalist Division of Labor." *Telos* 12 (Summer):27–43.

——. 1982. *Farewell to the Working Class.* London: Pluto Press.

Gossel, Patricia Peck. 1992. "A Need for Standard Methods: The Case of American Bacteriology." Pp. 287–311 in *The Right Tools for the Job: At Work in Twentieth-Century Life Sciences,* edited by Adele Clark and Joan Fujimura. Princeton: Princeton University Press.

Grayson, Lawrence P. 1993. *The Making of an Engineer.* New York: John Wiley.

Grinter, L. E. 1955. "Report of the Committee on Evaluation of Engineering Education." *Journal of Engineering Education,* 46:264–310.

Groopman, Leonard C. 1987. "Medical Internship as Moral Education: An Essay on the System of Training Physicians." *Culture, Medicine, and Psychiatry,* 11:207–27.

Gunz, Hugh P. 1980. "Dual Ladders in Research: A Paradoxical Organizational Fix." *R&D Management,* 10:113–18.

Haber, Samuel. 1991. *The Quest for Authority and Honor in the American Professions 1750–1900.* Chicago: University of Chicago Press.

Hall, Richard H. 1968. "Professionalization and Bureaucratization." *American Sociological Review,* 33:92–104.

——. 1986. *Dimensions of Work.* Beverly Hills: Sage.

Hall, Robert T. 1993. "Introduction." Pp. 11–53 in *E. Durkheim Ethics and the Sociology of Morals.* Buffalo, N.Y.: Prometheus Books.

Harper, Douglas. 1987. *Working Knowledge: Skill and Community in a Small Shop.* Chicago: University of Chicago Press.

Harries-Jenkins, G. 1970. "Professionals in Organizations." Pp. 51–107 in *Professions and Professionalization,* edited by J. A. Jackson. Cambridge: Cambridge University Press.

Held, Marilyn S. 1991. "Project Imagelift: Membership Has Its Benefits." *Laboratory Medicine,* 29:601–2.

Henderson, Kathryn. 1991. "Flexible Sketches and Inflexible Data Bases: Visual Communication, Conscription Devices and Boundary Objects in Design Engineering." *Science, Technology & Human Values,* 16:448–73.

Hirschhorn, Larry. 1984. *Beyond Mechanization.* Cambridge, Mass.: MIT Press.

Hochschild, Arlie. 1989. *The Second Shift.* New York: Viking.

Hodson, Randy, Gregory Hooks, and Sabine Rieble. 1992. "Customized Training in the Workplace." *Work and Occupations,* 19:272–92.

Hoffmaster, Barry. 1990. "Morality and the Social Sciences." Pp. 241–60 in *Social Science Perspectives on Medical Ethics,* edited by George Weisz. Boston, Mass.: Kluwer Academic Publishers.

Holzner, Burkart, and John Marx. 1979. *Knowledge Application: The Knowledge System in Society.* Boston: Allyn and Bacon.

Hounshell, David, and John Smith. 1988. *Science and Corporate Strategy: DuPont R&D, 1902–1980.* New York: Cambridge University Press.

Howell, Eleanor, Linda Wildra, and M. Gail Hill. 1988. *Comprehensive Trauma Nursing: Theory and Practice.* Glenview, Ill.: Chandler Press.

Hughes, Everett C. 1958. *Men and Their Work.* Glencoe, Ill.: The Free Press.

Hull, Daniel M. 1986. "Introduction: Preparing Technicians for Tomorrow's Jobs" Pp. vii–xvii in *Technician Education Directory.* Ann Arbor, Mich.: Prakken Publications.

Institute of Society, Ethics, and the Life Sciences. 1978. *In the Service of the State: The Psychiatrist as Double Agent.* Hastings Center Report, Special Supplement.

Jackall, Robert. 1988 *Moral Mazes.* New York: Oxford University Press.

Jordon, Kathleen, and Michael Lynch. 1992. "The Sociology of a Genetic Engineering Technique: Ritual and Rationality in the Performance of the 'Plasmid Prep.'" Pp. 77–114 in *The Right Tools for the Job: At Work in Twentieth-Century Life Sciences,* edited by Adele E. Clarke and Joan H. Fujimura. Princeton: Princeton University Press.

Katz, Ralph, ed. 1988. *Managing Professionals and Innovative Organizations.* Englewood Cliffs, N.J.: Prentice-Hall.

Katz, Ralph, Michael L. Tushman, and Thomas J. Allen. 1992. "Managing the Dual Ladder: A Longitudinal Study." Pp. 133–50 in *Advances in the Management of Technology,* vol. 1, edited by Luis Gomez-Mejia and Michael W. Lawless. Greenwich, Conn.: JAI Press.

Kern, Horst, and Michael Schumman. 1992. "New Concepts of Production and the Emergence of the Systems Controller." Pp. 111–48 in *Technology and the Future of Work,* edited by Paul Adler. New York: Oxford University Press.

Kidder, Tracy. 1981. *The Soul of a New Machine.* Boston: Little, Brown and Company.

Klein, D. P. 1984. "Occupational Employment Statistics: 1972–82." *Employment and Earnings,* 20:13–16.

Knight, Mary. 1991. "The Histology Professional: A Special Breed." *Laboratory Medicine,* 12:843–44.

Knorr-Cetina, Karin D. 1981. *The Manufacture of Knowledge: An Essay on the Constructivist and Contextual Nature of Science.* Oxford: Pergamon.

Koch, James. 1977. "Status Inconsistency and the Technician's Work Adjustment." *Journal of Occupations Psychology,* 50:121–28.

Kocka, Jürgen. 1980. *White Collar Workers in America 1890–1940.* Beverly Hills, Calif.: Sage.

Kokklenberg, Edward, and Donna Sockell. 1985. "Union Membership in the United States, 1973–1981." *Industrial and Labor Relations Review,* 38:497–543.

Kornhauser, William. 1962. *Scientists in Industry: Conflict and Accommodation.* Berkeley: University of California Press.

Kuhn, Sarah, and Charles Richardson. 1993. "Standing on the Edge: Engineering Students Encounter Skill-Based Automation." Pp. 19–26 in *Automated Systems Based on Human Skill and Intelligence,* edited by N. Harvey and Frank Emspack. N.Y.: Pergamon Press.

Kuhn, T. S. 1961. "The Function of Measurement in Modern Physical Science," *Isis.* 52:161–90.

Kunda, Gideon. 1992. *Engineering Culture: Control and Commitment in a High-Tech Corporation.* Philadelphia: Temple University Press.

Kurtz, Ronald, and Goodyear K. Walker. 1975. "Test and Pay Technicians—Upward Mobility in Personnel." *Public Personnel Management,* July–August: 259–62.

Kusterer, Ken C. 1978. *Know-how on the Job: The Important Working Knowledge of Unskilled Workers.* Boulder, Colo.: Westview.

Larkin, Gerald V. 1983. *Occupational Monopoly and Modern Medicine*. London: Tavistock.

Larson, Magali Sarfatti. 1977. *The Rise of Professionalism: A Sociological Analysis*. Berkeley: University of California Press.

Latour, Bruno. 1987. *Science in Action: How to Follow Scientists and Engineers through Society*. Cambridge, Mass.: Harvard University Press.

Latour, Bruno, and Steve Woolgar. 1979. *Laboratory Life: The Construction of Scientific Facts*. Princeton: Princeton University Press.

Lave, Jean, and Etienne Wenger. 1991. *Situated Learning: Legitimate Peripheral Participation*. Cambridge: Cambridge University Press.

Layton, Edwin. 1971. *The Revolt of the Engineers: Social Responsibility and the American Engineering Profession*. Baltimore, Md.: Johns Hopkins Press.

Lee, Gloria L., and Chris Smith. 1992. *Engineers and Management: International Comparisons*. London: Routledge.

Lévi-Strauss, Claude. 1966. *The Savage Mind*. Chicago: Chicago University Press.

Levitan, Sar A. 1984. "The Changing Workplace." *Society*, 216:41–48.

Lynch, M. 1985. *Art and Artifact in Laboratory Science: A Study of Shop Work and Shop Talk in a Research Laboratory*. London: Routledge and Kegan Paul.

Machlup, Fritz. 1980. *Knowledge: Its Creation, Distribution, and Economic Significance*. Vol. 1. Princeton, NJ: Princeton University Press.

MacKenzie, Donald, and Graham Spinardi. 1995. "Tacit Knowledge, Weapons Design, and the Uninvention of Nuclear Weapons." American Journal of Sociology, 101:44–99.

Majchrzak, Ann. 1988. *The Human Side of Factory Automation*. San Francisco: Jossey Bass.

Mallet, Serge. 1975. *The New Working Class*. Nottingham: Spokesman Books.

Marcson, Simon. 1960. *The Scientist in American Industry: Some Organizational Determinants in Manpower Utilization*. Princeton: Princeton University.

Marcuse, Herbert. 1968. *L'homme Unidimensionnel*. Paris: Editions de Minuit.

May, William F. 1983. *The Physician's Covenant*. Philadelphia: Westminster Press.

McCallion, R. 1987. "Paramedic Scope of Practice under Fire." *The Protocall*, San Francisco Paramedic Association, Calif.: June/July.

McIlwee, Judith A., and J. Gregg Robinson. 1992. *Women in Engineering: Gender, Power, and Workplace Culture*. Albany, N.Y.: State University of New York Press.

McQueen, Iris, and James L. Paturas. 1990. "EMS at Work in the 90s." *Emergency*, September: 38–41.

Meiksins, Peter. 1989. "Engineers and Managers: An Historical Perspective on an Uneasy Relationship." Paper presented at the American Sociological Association Annual Meetings, San Francisco.

Metz, Donald L. 1981. *Running Hot: Structure and Stress in Ambulance Work*. Cambridge, Mass.: Abt Publishers.

Midgley, Mary. 1992. *Science as Salvation*. New York: Routledge.

Miller, A., D. J. Treiman, P. S. Cain, and P. A. Roos. 1980. *Jobs and Occupations: A Critical Review of the Dictionary of Occupational Titles*. Washington, D.C.: National Academy Press.

Mills, C. Wright. 1956. *White Collar: The American Middle Classes*. New York: Oxford.

Mulkay, Michael J. 1984. "Knowledge and Utility: Implications for the Sociology of Knowledge." Pp. 77–98 in *Society and Knowledge: Contemporary Perspectives on the Sociology of Knowledge*, edited by N. Stehr and V. Mega. New Brunswick: Transaction Books.

Murray, Thomas H. 1986. "Divided Loyalties for Physicians: Social Context and Moral Problems." *Social Science and Medicine*, 238:827–32.

National Academy of Sciences/National Research Council. 1966. *Accidental Death and Disability.* Washington, D.C.: National Academy of Sciences.
National Society of Professional Engineers. 1992. *Engineering Education Issues: Report on Surveys of Opinions by Engineering Deans and Employers of Engineering Graduates on the First Professional Degree.* Publication number 3059. Alexandria, Va.: National Society of Professional Engineers.
Nelsen, Bonalyn J., and Stephen R. Barley. "The Social Negotiation of a Skilled Occupational Identity." Working paper, Center for the Educational Quality of the Workforce, University of Pennsylvania, Philadelphia, 1994.
——. "Toward an Emic Understanding of Professionalism among Technical Workers." Working paper, Center for the Educational Quality of the Workforce, University of Pennsylvania, Philadelphia, 1994.
Nelson, Daniel. 1975. *Origins of the New Factory System in the United States: 1880–1920.* Madison, Wisc.: University of Wisconsin Press.
Ng, Sek-Hong. 1986. "Electronics Technicians in an Industrializing Economy: Some Glimpses of the 'New Middle Class.'" *Sociological Review,* 34:611–40.
Noble, David F. 1977. *America by Design: Science, Technology, and the Rise of Corporate Capitalism.* New York: Oxford.
——. 1984. *Forces of Production: A Social History of Industrial Automation.* New York: Knopf.
Orr, Julian E. 1990. "Sharing Knowledge, Celebrating Identity: Community Memory in a Service Culture." Pp. 169–89 in *Collective Remembering,* edited by D. Middleton and D. Edwards. Newberry Park, Calif.: Sage.
——. 1991a. "Necessary or Prescribed Practices: An Ethnographic Approach to Conflicting Definitions of Work and the Realities of the Job." Palo Alto, Calif.: Systems Sciences Laboratory, Xerox PARC.
——. 1991b. "Contested Knowledge." *Anthropology of Work Review,* 12:12–17.
——. 1996. *Talking about Machines: An Ethnography of a Modern Job.* Ithaca, NY: ILR Press.
Palmer, C. Eddie, and Sheryl M. Gonsoulin. 1990. "Paramedics, Protocols, and Procedures: 'Playing Doc' as Deviant Role Performance." *Deviant Behavior,* 11:207–19.
Partridge, Eric. 1966. *Origins: A Short Etymological Dictionary of the English Language.* New York: Macmillan.
Pelz, Donald C., and Frank M. Andrews. 1966. *Scientists in Organizations: Productive Climates for Research and Development.* Ann Arbor, Mich.: Institute for Social Research, The University of Michigan.
Pentland, Brian T. 1991a. "Making the Right Moves: Toward a Social Grammar of Software Support Hot Lines." Diss., Massachusetts Institute of Technology.
——. 1991b. "Organizing Moves in Software Support Hot Lines." *Administrative Science Quarterly,* 37:527–48.
——. In press. "Read Me What It Says on Your Screen: The Interpretive Problem in Technical Service Work." *Technology Studies.*
Pepe, Paul E., and Marni J. Bonnin. 1989. "Limitations and Liabilities." *Emergency,* March: 40–43.
Perlow, Leslie. 1993. "The Myth of 'Real Work': A Case Study of Engineers' Preferences and the Organization's Needs." Sloan School of Management, MIT.
——. 1994. "Putting the Work Back into Work/Family." Sloan School of Management, MIT.
——. 1995. "Time Famine: The Unintended Consequence of the Definition of Success." Diss., Massachusetts Institute of Technology.

Perrow, Charles. 1984. *Normal Accidents: Living with High Risk Technologies*. New York: Basic Books.

Perrucci, Robert. 1969. *Engineers and the Social System*. New York: Wiley.

Perrucci, Robert, and Joel E. Gerstl. 1969. *Profession without Community: Engineers in American Society*. New York: Random House.

Peterson, Nancy. 1988. "The Black and White World of Private EMS." *Journal of Emergency Medical Services*, October: 38–43, 68–69.

Pettigrew, Andrew M. 1973. *The Politics of Organizational Decision Making*. London: Tavistock.

Piore, Michael J., and Charles F. Sabel. 1984. *The Second Industrial Divide*. New York: Basic Books.

Porter, Roy. 1977. *The Making of Geology: Earth Science in Britain 1660–1815*. Cambridge: Cambridge University Press.

Poulantzas, Nicos. 1975. *Classes in Contemporary Capitalism*. London: New Left Books.

Price, Dick. 1993. "Cows and Computers." *IEEE Micro*, December: 95–97.

Raelin, Joseph A. 1985. *The Clash of Cultures: Managers and Professionals*. Boston: Harvard Business School Press.

Reich, Leonard S. 1985. *The Making of American Industrial Research: Science and Business at GE and Bell, 1876–1926*. NY: Cambridge University Press.

Richman, Louis S. 1994. "The Worker Elite." *Fortune*, August 22, Pp. 56–66.

Ritti, R. Richard. 1971. *The Engineer in the Industrial Corporation*. New York: Columbia University Press.

——. 1982. "Work Goals of Scientists and Engineers." Pp. 363–75 in *Readings in the Management of Innovation*, edited by Michael L. Tushman and William L. Moore. Cambridge, Mass.: Ballinger Publishing Company.

Roberts, B. C., Ray Loveridge, John Gennard, and J. V. Eason. 1972. *Reluctant Militants: A Study of Industrial Technicians*. London: Heinemann Educational Books.

Roberts, Edward B., and Alan R. Fusfeld. 1982. "Critical Functions: Needed Roles in the Innovation Process." Pp. 182–207 in *Career Issues for Human Resource Management*, edited by Ralph Katz. Englewood Cliffs, N.J.: Prentice-Hall.

Roberts, H. 1983. "A Qualified Failure." *New Scientist*, June 9, p. 722.

Robinson, J. Gregg, and Judith S. McIlwee. 1991. "Men, Women, and the Culture of Engineering." *Sociological Quarterly*, 32:403–21.

Rosenberg, S. A. 1991. "Hospitals Grapple with HCFA's Rules for Clinical Labs." *Trustee*, June: 18–19.

Scarselletta, Mario. "Button Pushers and Ribbon Cutters: Observations on Skill and Practice in a Hospital Laboratory and Their Implications for the Shortage of Skilled Technicians." Working paper, Center for the Educational Quality of the Workforce, University of Pennsylvania, Philadelphia, 1992.

Schafft, Gretchen E., and James F. Cawley. 1987. *The Physician's Assistant in a Changing Health Care Environment*. Rockville, Md.: Aspen Publishers.

Schon, Donald A. 1988. *Educating the Reflective Practitioner*. New York: Basic Books.

Schriesheim, Janet, Mary Ann Von Glinow, and Steven Kerr. 1977. "Professionals in Bureaucracy: A Structural Alternative." *Management Science*, 5:55–69.

Schumpeter, J. A. 1942. *Capitalism, Socialism and Democracy*. New York: Harper and Row.

Seely, Bruce E. 1993. "Research, Education, and Science in American Engineering Colleges, 1900–1960." *Technology and Culture*, 34:344–86.

Sennett, Richard, and Jonathan Cobb. 1972. *The Hidden Injuries of Class*. New York: Knopf.

Shapin, Steven. 1989. "The Invisible Technician." *American Scientist,* 7:554–63.
Sharp, Art, and Betsy Sharp. 1989. "Times Are A'Changin." *Emergency,* March: 28–30.
Silvestri, George, and John Lukasiewicz. 1991. "Occupational Employment Projections." *Monthly Labor Review,* Winter: 64–94.
Simpson, Richard. 1985. "Social Control of Occupations and Work." *Annual Review of Sociology,* vol. 11, pp. 415–36.
Slichter, Sumner, James Healy, and E. Robert Livernash. 1960. *Impact of Collective Bargaining on Management.* Washington: Brookings Institution.
Smith, Chris. 1987. *Technical Workers: Class, Labour, and Trade Unionism.* London: Macmillan.
Smith, Chris, and Peter Whalley. 1996. "Engineers in Britain: A Study in Persistence." Forthcoming in *Engineering Class Politics,* edited by Peter Meiksins and Chris Smith. London: Verso.
Smith, Deborah Takiff. 1984. *Computers on the Farm.* U.S. Department of Agriculture Farmer's Bulletin no. 2277. Washington: U.S. Department of Agriculture.
Smith, J. P., and B. I. Bodai. 1983. "Prehospital Stabilization of Critically Injured Patients: A Failed Concept." Presentation at the 43rd Annual Meeting of the American Association for the Surgery of Trauma, October.
——. 1985. "The Urban Paramedic's Scope of Practice." *JAMA,* 2534:544–48.
Sonenscher, Michael. 1987. "Mythical Work: Workshop Production and the Compagnonnages of Eighteenth-Century France." Pp. 31–63 in *The Historical Meanings of Work,* edited by Patrick Joyce. New York: Cambridge.
Spector, Bert A. 1982. "Air Traffic Controllers." Cambridge, Mass.: Harvard Business School Case 9-482-056.
Spenner, Kenneth I. 1980. "Occupational Characteristics and Classification Systems: New Uses of the Dictionary of Occupational Titles." *Sociological Methods and Research,* 9:239–64.
——. 1991. "Skill: Meanings, Methods, and Measures." *Work and Occupations,* 17:399–421.
Stabile, Donald. 1984. *Prophets of Order: The Rise of the New Class, Technocracy and Socialism in America.* Boston: South End Press.
Stevens, Gillian, and Cho Joo Hyun. 1985. "Socioeconomic Indexes and the New 1980 Census Occupational Classification Scheme." *Social Science Research,* 14:142–68.
Stinchcombe, Arthur. 1959. "Bureaucratic and Craft Administration of Production: A Comparative Study." *Administrative Science Quarterly,* 4:168–87.
Streett, William B. 1993. "The Military Influence on American Engineering Education." *Cornell Engineering Quarterly,* vol. 27: 2 (Winter).
Sturt, George. 1923. *The Wheelwright's Shop.* Cambridge: Cambridge University Press.
Szafran, Robert F. 1992. "Measuring Occupational Change over Four Decennial Censuses, 1950–1980." *Work and Occupations,* 19:293–327.
Taylor, Frederick Winslow. 1911. *The Principles of Scientific Management.* New York: Norton.
Thomas, Robert J. 1994. *What Machines Can't Do: Politics and Technology in the Industrial Enterprise.* Berkeley: University of California Press.
Toulmin, Stephen. 1986. "Divided Loyalties and Ambiguous Relationships." *Social Science and Medicine,* 238:783–87.
Touraine, Alain. 1971. *The Post-Industrial Society.* New York: Random House.
Travis, G. D. L. 1981. "Replicating Replication? Aspects of the Social Construction of Learning in Planarian Worms." *Social Studies of Science,* 11:11–32.

Trice, Harrison M. 1993. *Occupational Subcultures in the Workplace.* Ithaca, N.Y.: ILR Press.
Turkle, Sherry. 1984. *The Second Self: Computers and the Human Spirit.* New York: Simon and Schuster.
U.S. Bureau of the Census. 1976. *The Statistical History of the United States from Colonial Times to the Present.* New York: Basic Books.
——. 1982. *1980 Census Occupational Classification.* Washington: U.S. Government Printing Office.
U.S. Department of Commerce. 1991. *Statistical Abstract of the United States.* Washington: U.S. Government Printing Office.
U.S. Department of Labor. 1977. *Dictionary of Occupational Titles,* 4th ed. Washington: U.S. Government Printing Office.
U.S. Department of Labor, Bureau of Labor Statistics. 1989. *Handbook of Labor Statistics.* Bulletin 2340, August. Washington: U.S. Government Printing Office.
——. 1994. *Employment and Earnings.* Washington: U.S. Government Printing Office.
Udy, Stanley, 1980. "The Configuration of Occupational Structure." Pp. 156–65 in *Sociological Theory and Research,* edited by Hubert Blalock, Jr. Glencoe, Ill: The Free Press.
Vallas, Peter. 1990. "The Concept of Skill: A Critical Review." *Work and Occupations,* 17:379–98.
Van Maanen, John, and Stephen R. Barley. 1984. "Occupational Communities: Culture and Control in Organizations." Pp. 287–365 in *Research in Organizational Behavior,* vol. 6, edited by Barry Staw and Larry Cummings. Greenwich, Conn.: JAI Press.
Vincenti, W. G. 1984. "Technological Knowledge without Science: The Innovation of Flush Riveting in American Airplanes, ca. 1930–ca. 1950." *Technology and Culture* 25:540–76.
Von Glinow, Mary Ann. 1988. *The New Professionals: Managing Today's High-Tech Employees.* Cambridge, Mass.: Ballinger.
von Hippel, E. 1994. "Sticky Information and the Locus of Problem Solving: Implications for Innovation." *Management Science,* 40:429–39.
von Hippel, E., and M. Tyre. 1993. "How 'Learning by Doing' is Done: Problem Identification in Novel Process Equipment," Working paper #BPS 3521-93. Cambridge, Mass.: Sloan School of Management, MIT.
Watson, J. M., and P. F. Meiksins. 1991. "What Do Engineers Want? Work Values, Job Rewards and Job Satisfaction." *Science, Technology, & Human Values,* 16:140–72.
Weber, Max. 1968 [1922]. *Economy and Society.* New York: Bedminster Press.
Weiss, J. H. 1982. *The Making of Technological Man: The Social Origins of French Engineering Education.* Cambridge, Mass.: MIT Press.
Westergaard, John. 1984. "Class of 84." *New Socialist,* 15:30–6.
Westwood, Sallie. 1984. *All Day, Every Day: Factory and Family in the Making of Women's Lives.* London: Pluto.
Whalley, Peter. 1986a. "Markets, Managers, and Technical Autonomy." *Theory and Society* 15:223–47.
——. 1986b. *The Social Production of Technical Work.* Albany: State University of New York Press.
——. 1987. "Constructing an Occupation: The Case of British Engineers." *Current Research on Occupations and Professions,* vol. 4, pp. 3–20. Greenwich, Conn.: JAI Press.
——. 1990. "Markets, Managers and Technical Autonomy in British Plants." Pp. 373–394 in *Structures of Capital: The Social Organization of Economy,* edited by Sharon Zukin and Paul DiMaggio. Cambridge: Cambridge University Press.

——. 1991a. "Negotiating the Boundaries of Engineering: Professionals, Managers and Manual Work." *Research in the Sociology of Organizations,* 8:191–215.
——. 1991b. "The Social Practice of Independent Inventing," *Science, Technology and Human Values.* 16:208–32.
Whalley, Peter, and Stephen Crawford. 1984. "Locating Technical Workers in the Class Structure." *Politics and Society,* 13:239–52.
Wiener, Martin J. 1981. *English Culture and the Decline of the Industrial Spirit, 1850–1980.* Cambridge: Cambridge University Press.
Williamson, Oliver. 1975. *Markets and Hierarchies.* New York: The Free Press.
Wright, Erik O. 1978. *Class, Crisis, and the State.* London: New Left Books.
——. 1985. *Classes.* London: Verso.
Wright, James D. 1978. "In Search of a New Working Class." *Qualitative Sociology,* 1:33–57.
Wrigley, Julia. 1986. "Technical Education and Industry in the Nineteenth Century." Pp. 162–88 in *The Decline of the British Economy,* edited by B. Elbaum and W. Lazonick. Oxford: Oxford University Press.
Wuthnow, Sara. 1986. "Nurses and the New Class." *Sociological Inquiry,* 56:125–48.
Yount, Kristen R. 1991. "Ladies, Flirts, and Tomboys: Strategies for Managing Sexual Harassment in an Underground Coalmine." *Journal of Contemporary Ethnography,* 19:396–422.
Zabusky, Stacia E. "Strain and the Organizational Scientist: A Cultural Explanation." Working paper, Center for the Educational Quality of the Workforce, University of Pennsylvania, Philadelphia, 1993.
Zabusky, Stacia E., and Stephen R. Barley. 1993. Fieldnotes from a study of microcomputer support technicians. Program on Technology and Work, Cornell University, Ithaca, N.Y.
——. 1994. "Liminality and the Organizational Scientist: A Re-evaluation of Professional Identities in Organizations." Working paper, Program on Technology and Work, Cornell University, Ithaca, N.Y.
——. 1996. "Redefining Success: Ethnographic Observations on the Careers of Technicians." Pp. 185–214. In Paul Osterman (ed.), *Broken Ladders: White Collar Careers.* Oxford: Oxford University Press.
Zimbalist, Andrew. 1979. *Case Studies in the Labor Process.* New York: Monthly Review Press.
Zuboff, Shoshana. 1988. *In the Age of the Smart Machine.* NY: Basic Books.
Zussman, Robert. 1985. *Mechanics of the Middle Class.* Berkeley: University of California Press.

# INDEX